PIONEERS
of
ROCKETRY

PIONEERS
of
ROCKETRY

Michael Stoiko

Introduction by
Frederick C. Durant III

HAWTHORN BOOKS, Inc.
Publishers / New York

Library of Congress Catalog Card Number: 73-14229

ISBN: 0-8015-5876-X

1 2 3 4 5 6 7 8 9 10

*To my brother Ed, who
dreamed along with me*

Contents

Foreword

Those who pioneer in fields of science and technology before such fields are accepted by their peers have not an easy lot. Certain personality characteristics are required to advance in the face of disinterest, lack of appreciation and, often, ridicule. Qualities of imagination, perseverance, conviction, and faith are needed—together with the capability for painstaking detail and plain hard work. Progress is made only when discouragement will not be accepted.

This has certainly been true of the five men whose lives and accomplishments are related in this book. Each had insight into the many problems involved, but also the enormous satisfaction of discovery. Except for Hermann Oberth, recognition was belated. Professor Oberth, alone among the twentieth-century pioneers, has lived to see man on the moon.

William Congreve was an empirical technologist who developed the crude oriental rocket into a powerful military system. Konstantin Eduardovich Tsiolkovsky was a mathematician and creative visionary who invented—on paper—rocket-powered space ships. Before the nineteenth century he had begun to explore the multifaceted and then unknown field of astronautics, contributing concepts and fundamental mathematical formulas that were to become practical decades later. Tsiolkovsky was an inspiration to men who would later produce Soviet space accomplishments. Hermann Oberth, similarly, inspired youthful engineers and scientists who later created triumphs in German rocket technology and United States space programs.

Robert Esnault-Pelterie and Robert Hutchings Goddard were both dreamers and inventors. Both developed theories of rocket propulsion and space flight. Both designed and developed liquid-propellant rocket motors. Dr. Goddard, however, conducted scientific investigations of liquid-propellant rockets earlier—and for a longer time—than any of the others. Further, he was the first to actually launch a liquid-propellant rocket. His recently published private papers include ideas and details for manned flight to the moon, the planets, and the stars. These papers date back more than fifty years.

Michael Stoiko has assembled the life stories of these rocket and space pioneers in an interesting fashion. It is quite probable that some of the young men and women who read this work will have the urge to blaze their own trail in scientific and technical exploration and discovery. If so, they too may expect hard work, probable disappointments and, hopefully, the thrill of accomplishment known by the men in this book.

FREDERICK C. DURANT III

Preface

Rocketry, in a very real sense, is the product of the international genius of mankind. In general, it is the brainchild of the pioneering rocket greats; and, in particular, it is synonymous with such names as Congreve, Tsiolkovsky, Esnault-Pelterie, Goddard, and Oberth. *Pioneers of Rocketry* is their story. It is a story of the men who have helped alter the course of human history and have taken us in a new direction.

The preparation of this book required the help of many friends and organizations. In particular I wish to express my gratitude to Esther C. (Mrs. Robert H.) Goddard for her cooperation and permission to use her husband's pictures and materials; Dr. José Martinez for his translations from *L'Astronautique;* Dr. Hermann Oberth for his fine letter; Lise Blosset of the French Centre National d'Études Spatiales (CNES) for her help in obtaining biographical data on Esnault-Pelterie; the Soviet embassy for its help with Tsiolkovsky; Frank Winter of The Smithsonian Institution for his knowledgeable help in researching materials; George James, National Science Foundation, for his aid in obtaining biographical data on Esnault-Pelterie; John Bell of the British embassy for his Congreve data; Lee D. Saegesser of the Historical Branch of the National Aeronautics and Space Administration (NASA) for his very helpful cooperation; and, as always, The Smithsonian Institution and NASA for providing leads, research material, and photographs that are used throughout this book.

My special thanks to Frederick C. Durant III, assistant director, National Air and Space Museum, The Smithsonian Institution, and old friend, for his invaluable assistance and for writing the Introduction to *Pioneers of Rocketry*.

PIONEERS
of
ROCKETRY

1

Ancient Rocket Pioneers

> I sometimes think the desire to fly . . .
> is an idea handed down to us by our
> ancestors who, in their grueling travels
> across trackless lands in prehistoric
> times, looked enviously on the birds
> soaring freely through space . . . on
> the infinite highway of the air.
> Wilbur Wright, 1908

The desire to fly has been felt by man throughout recorded history. We find this interest imaginatively expressed in several forms.

Some of the earliest signs of interest in flying would be symbolic animals like the winged bulls guarding the ancient halls of Persia or the winged horse, Pegasus, of Greek mythology. Besides creating imaginary animals with wings, ancient legends also tell of man's attempts to become airborne.

One of these stories was written about 4,000 years ago in the Chinese book *Annals of the Bamboo Books*. The story is is about the Emperor Shun, who as a boy escaped from captivity "by donning the work-cloths of a bird." On still another occasion he became a flying dragon. Other Chinese legends tell of Lei Kung, the god of thunder and lightning, who had wings of a bat, and of Ki-kung-shi, who in the first century B.C. invented a flying chariot.

The legends from almost every early civilization have in-

3

cluded stories of animal or manned flights. The ancient king of Persia, Kai Kawus, is credited with having a flying throne that was carried by four eagles; other legends say Alexander the Great had a cage that could be carried by winged griffins (a mythical animal with the body and hind legs of a lion and the head and wings of an eagle).

Perhaps the best known of all the flying legends is the story of Daedalus and Icarus who constructed wings of wax and feathers so that they could escape from the Cretan King Minos; as we all know, the flight was aborted when Icarus flew too close to the sun. Jules Verne believed along with many others that in fashioning the wings that killed his son Daedalus was attempting to build a glider large enough to carry a man. If this was the case, then Icarus was the first fatality in man's quest to fly.

In time, legend gave way to records of man's actual attempts to build objects that fly. One of the earliest of the more successful attempts is said to have occurred several centuries before the birth of Christ. Kung-shu Tse, a contemporary of Confucius, was supposed to have built a bamboo magpie that flew for three days. The magpie was a kite. By the third century B.C. kites were being used for military signaling, and by the Middle Ages kites had been developed large enough to lift a man.

We find an account from Greece of about 400 B.C. concerning the first applications of the reaction principle. This is the story of the mysterious wooden pigeon of Archytas, which astounded observers in southern Italy. One report on the flight claims that the pigeon was able to fly by means of "hidden and enclosed air" and that it was controlled by a string or rod during its flight, much like the captive flights of model airplanes today. In any event, it is generally accepted that the wooden pigeon was made to fly by means of a reaction jet of steam or compressed air.

The ancient pioneers continued their efforts to perfect

manned flight despite the tremendous odds they faced. For hundreds of years arm-powered wings were the only hope they had, launched from cliffs and later from high towers.

High towers attracted many winged jumpers according to legend and history. One of the better known was Bladud, the legendary tenth king of Britain. According to medieval sources, Bladud was killed attempting to fly over London with a pair of wings. (Bladud was also the father of King Lear, which suggests that madness probably ran in the family.)

Another story from Greece, dated about A.D. 160, concerns the Greek mathematician Heron of Alexandria. Heron is given credit for inventing the aeolipile. The aeolipile, named after Aeolus, the god of the wind, was the first known mechanical device to demonstrate the reaction principle by converting steam pressure into jet reaction. Similar devices are still used in physics classes to demonstrate the reaction principle.

Its operation was simple. A pot with a cover and legs was filled with water, and a fire was then built under the pot. The aeolipile was a hollow sphere mounted between two tubular supports in such a manner that the sphere was free to rotate. The same tubular supports were also used to carry steam into the sphere.

The aeolipile also had two right-angle pipes located opposite each other. These pipes served as the exhaust nozzles. As the flame heated the water in the pot, it produced steam. Because of the pressure, the steam traveled from the pot, through the hollow pipe, into the sphere. In time, the pressure built up in the sphere, and soon the steam was forced through the second set of pipes and began to spout from the nozzles. The steam jet spouting out of the nozzle pipes soon produced enough pressure, or thrust as it is called in rocketry, to spin the sphere around faster and faster, like a pinwheel. This spinning continued as long as steam was produced by the pot.

It is easy to see that Heron's invention was the basic idea

that in time developed into the steam engine, the steam turbine, the jet-propelled airplane, and the modern rocket. Probably one of the best-known simple applications of this idea is the rotating lawn sprinkler. Another more-complicated use of this idea is that of the steam reaction jet used for flight control in rockets, rocket aircraft, and all spacecraft. In fact, this very idea was first used in the Viking rocket, later in the X-15 airplane, and then in all modern spacecraft—like the Mercury, Gemini, and Apollo. So we see how a simple idea was improved on through seventeen centuries and handed down to us for use in modern rocketry and space flight. The aeolipile was doomed to remain a toy for scholars in Heron's day, however, for there was no practical application of this idea.

Another contribution to rocketry made by the Greeks in A.D. 673 was the discovery of a rapidly burning flash-powder mixture made up of pitch and naphtha. Known as "Greek fire," it was poured down from battlements onto invading troops. After its first use by the Greeks news of the invention spread far and wide, perhaps even to China.

This may have been one of the ways the Chinese acquired their knowledge of powder rocketry. Still another way the Chinese may have discovered it was to notice that saltpeter, which was used for curing meat, burned with a flash when dropped into a fire. Saltpeter and charcoal were the two chemicals used to make the powder known as "Chinese fire." It is assumed that by adding sulfur to this mixture the Chinese eventually discovered gunpowder.

One of the first recorded uses of gunpowder and rockets came more than 500 years later. In an ancient and well-illustrated Chinese manuscript known as *T-hung-lian-kang-mu* the story of a rocketlike device is told. A great siege was laid to the South China city of Kai-fung-fu (Pien-ching) in 1232 by Mongol hordes led by Ogdai, son of Genghis Khan.

The manuscript tells how during this battle two new uses

火籠箭式

A drawing of Chinese fire rockets by Wu-Pei-Chih (ca. 1620).
(*The Smithsonian Institution*)

of gunpowder weapons were made by the defending Chinese. The first was called Arrow of Flying Fire, and the second was known as Heaven-Shaking Thunder. The story tells how Heaven-Shaking Thunder were bombs dropped on the invaders from the battlements above and that the Arrow of Flying Fire flew out several paces. The Arrow of Flying Fire, however, may also have been a fire-scattering device such as a hand-held Roman candle. The description could fit either of these devices.

The Arrows of Flying Fire were launched under their own power and were not shot from bows. They also flew in a reasonably straight line at the attackers. The resulting fire also spread over an area some ten feet in diameter. Although these new weapons did make an impressive, loud, flaming display that probably frightened the Mongol invaders, they proved to be of little practical use, for Ogdai's hordes finally managed to capture and destroy the city.

After the Battle of Pien-king the Chinese continued to improve on the art of making and using gunpowder. About 1270 Marco Polo first saw the Chinese use this explosive mixture. He was deeply impressed and brought back to Europe the idea of gunpowder. Within the next 130 years, from 1270 to the beginning of the fifteenth century, the making of gunpowder and the use of rockets became well known throughout Europe and other parts of the world.

During the 1400s rockets really began to come into their own as weapons. Konrad Kyeser von Eichstadt, a German military engineer of the period, wrote about three different types of rocket warfare weapons: vertically rising skyrockets, floating rockets, and rockets running along stretched string for amusement or for carrying messages. Still another military expert, Joanes de Fontana, an Italian who lived at the same time, showed real imagination in design in a book about rockets. He designed rockets as rabbits, pigeons, and fish, equipped with rollers to carry them toward the enemy. De

Fontana even designed a rocket car mounted on rollers. The rocket propelled the car toward enemy lines, where it acted as a battering ram and crushed the opposing forces. In the latter part of the seventeenth century Sir Isaac Newton, the English scientist, designed a wagon that moved forward with the aid of high-pressure steam coming out of the back. At about the same time in Holland a Dutch professor named Jacob Willem s'Gravesande designed another steam reaction car, but there is some question about whether a full-size one was ever constructed. In his car, as in Newton's, high-pressure steam was forced out through a tail pipe mounted in the rear, propelling the car forward.

Rockets were used in many ways in warfare by 1450. They were used in several battles against horse cavalry, as well as to set fire to wooden and thatched structures, to frighten the enemy, to send messages between separate encampments of armies, and to light up the sky at night. There are also some reports that rockets were used by pirates in attacks on ships to set fire to the highly flammable riggings on sailing vessels.

It was during the fifteenth century that cannons and small firearms were being developed by most nations. As these were perfected, they replaced the rocket as an important military weapon. In Europe by the sixteenth century guns had been improved to such an extent that rockets were totally discarded as weapons, being used solely for amusement and as fireworks at celebrations.

In the meantime the Chinese reportedly had developed the rocket for signaling purposes as well as for military weapons. Legend has it that these still very primitive devices gave a Chinese servant named Wan Hu, who lived at the beginning of the sixteenth century, an interesting idea on how he might propel himself into space.

The inventive Wan Hu secured two large paper kites, one to each hand, hoping that they would help lower him back to earth after his successful flight. He then sat down on a chair to which he had attached forty-seven of the largest rockets he could buy. When all was ready, he signaled his assistants to light the rockets with long torches.

As the countdown got to T plus one, the assistants ran up to the chair; at T equals zero they applied the torches to all forty-seven of the rockets at one time; and at T plus one there was a great roar and a blast of flame and smoke that completely hid Wan Hu from view. When the smoke cleared at T plus five, the assistants found that Wan Hu had vanished. Wan Hu was probably the first man to attempt space flight by using rockets as a means of transportation, and he was never heard from again.

It is obvious that this first experiment with manned rocket flight into space was not a success, but the courageous Wan Hu must be given credit for his daring, his imaginative idea of using rocket power for manned flight, and his pioneering spirit to explore the unknown. His feat surely must have inspired future generations of rocketeers to meet the challenges of space flight.

The next contribution to rocketry came from Russia and involved Peter the Great (Tsar Peter I). Peter's biography states that he was interested in rocketry in his youth and that he experimented with fireworks that were "ingenious, home-made, but none the less dangerous."

There are a number of conflicting dates about when gunpowder and rockets were first used in Russia. These claims cover a period of about 300 years (1375–1680), and each claim quotes one or more documents in proof. Of one thing we are sure: Peter the Great is said to have founded the first "rocket works" in Moscow in 1680, where signal and illuminating rockets were made for the Russian army. Later, in the early 1700s, Peter moved the rocket works to his new capital in St. Petersburg and made it much larger.

It is of particular interest to note that the history of Soviet rocketry from Peter the Great can be traced quite accurately for more than the next 270 years. Men like Aleksander D. Zasyadko (1779–1837), Konstantin I. Konstantinov (1818–1871), Nikolai I. Kilbalchich (1853–1881), Fridrikh A. Tsander (1877–1933), and Sergei Korolyov (1907–1966) were some of the Russian pioneers who can be linked with the development of rocketry in the Soviet Union.

One of the earliest important milestones in rocketry was reached in 1687 when Sir Isaac Newton presented his three laws of motion. The third law, dealing with action and reaction, is the principle on which a rocket works. His law states, "For every action there is an equal and opposite reaction, and the two act along the same straight line." Although Newton had provided man with the explanation of how a rocket operates, very little use was made of this information until almost 100 years later.

As we have seen, the development of rocketry did not happen accidentally in someone's basement, nor did it happen overnight. Rocketry was the product of many inventive men from many different nations. William Congreve, an Englishman, was one of them.

His story is the first of five biographical sketches of the great pioneers. The second is of a Russian pioneer; the third a Frenchman; the fourth an American; and the fifth a German. Their biographies, spanning a relatively short period in the history of rocket development, tell how a handful of inspired men transformed the crude Arrow of Flying Fire into the basic machinery that carried man into space and how the mathematics they developed helped us to understand the principles of obtaining orbital and interplanetary flight.

2

William Congreve

It must be laid down as a maxim that the very essence and spirit of the Rocket System is the facility of firing a great number of rounds in a short time . . . that the rocket is a species of fixed ammunition which does not require ordnance to project.

William Congreve, 1814

Sir William Congreve, born on May 20, 1772, was the creator of the British war rocket. His father, Capt. (later Lt. Gen.) William Congreve, was comptroller of the Royal Laboratory at the Woolwich Arsenal, and young William's home was on the arsenal grounds, along the banks of the Thames River just outside London. His playground was the big common, a parklike area of public land, which faced the barracks of the Royal Artillery, close to the Royal Gun Factory, the Royal Carriage Department, and the Royal Dockyards. The Woolwich Arsenal was responsible for the development and improvement of all British military weapons and for the fireworks that were used for national celebrations and royal events.

Congreve spent his childhood at the only place in all of England where skyrockets and weapons were discussed, developed, and built by experts; and, without a doubt, he must have learned a great deal about them at an early age.

A painting made when William was about ten years old shows him with his father on the arsenal grounds. The elder Congreve is pictured in the full uniform of a Royal Artillery officer, watching a gun squad haul a heavy gun into position. Young William is shown trying to attract his father's attention to a small toy gun he had set in its rack position for firing at a very high angle. The artist was Baron Johann Zoffany, a very well-known painter in his day, who painted influential people, members of the London stage, and members of the nobility. Since the baron painted the Congreves, it can be assumed that they too were socially prominent. Moreover, the family was related to William Congreve, the famous English dramatist (1670–1729).

William was very well educated. He attended the Hackney School outside London for the lower grades, the Singlewell School in the county of Kent through high school, and Trinity College at Cambridge University for his bachelor's degree in 1793. Then, in the tradition of the day, he went to London to study law at Middle Temple. He made his home at Garden Court, where many lawyers lived, and shortly thereafter began to publish and edit a political newspaper. But his newspaper career came to an abrupt end when he published an item about a Lord Berkely, who sued. Congreve was prosecuted for libel and fined £1,000.

In 1793 England went to war against France, a national event that started Congreve on his career of building war rockets. France had overthrown its king, and then the revolutionary armies under Napoleon set out to conquer the continent. When France attacked Holland, the Dutch appealed to England for help against the invading French forces. England reluctantly went to its aid. What England originally thought would be a short war turned out to be a struggle that lasted for more than twenty years.

It was during this period of the Napoleonic Wars that Congreve decided to work on a military weapon that might help England defeat the armies of France. Congreve was not an engineer or a scientist, but he lived at a time when educated men understood certain mechanical principles, among them Newton's laws of motion, that had been unknown to earlier experimenters. And with his background and experiences at Woolwich Arsenal he had an additional advantage over his predecessors in the work he undertook.

To learn all he could about rockets and weapons he borrowed books from the arsenal; one of them contained a detailed account of the rockets used during the British-Indian Wars. *A Narrative of the Military Operations on the Coromandel Coast* by Innes Munroe, Esq., published in London in 1789, described how in 1761 an Indian rocket corps of 1,200 men led by Hydar Ali, Prince of Mysore, defeated the British forces in the battle of Panipat, India. The rockets they used were a vast improvement over all the earlier known rockets. The rocket case, or body, was made of an iron tube instead of wood. It was about eight inches long and one and one-half inches in diameter and had a spiked nose. An eight-foot bamboo pole was attached as a stabilizer to control the rocket in flight. These rockets may not have been very accurate, but they traveled more than one-half of a mile and caused much damage when used in volleys of 2,000 at a time, especially when used against cavalry.

When the Indians saw how successful the rockets were, Hydar Ali's son, Tippoo Sahib, enlarged the rocket corps to 5,000 men, supplied them with still larger rockets, and defeated the British in battles several times from 1780 to 1799 near Seringapatam, India.

When Congreve felt that he had learned all there was to learn about skyrockets, guns, and gunpowder from the books available at the arsenal archives, he set out to conduct his first series of experiments with the skyrocket. He started by buying the largest fireworks that he could find in London. Some say he paid for them with his own money. Others claim that the

Admiralty provided the necessary funds for his experiments.

The skyrockets he bought were very similar to the ones made several centuries earlier. The cylindrical case was made of rolled hard paper, squeezed together at the bottom to form a narrow neck and then gradually opening up again to form a nozzle (much like our rocket motors). The front end of the cylinder carried the payload, consisting of powders specially mixed to provide exciting displays.

Congreve used the first set of rockets to find the proper angle at which to launch them; the second series of tests were conducted to find out the range of the skyrocket. Congreve soon learned that the skyrockets he had purchased could travel only about 500 yards, or less than one-half of the range of the Indian rockets. He next proceeded to improve the range of the skyrocket. By one or more of his various plans Congreve gradually increased the range of the paper board rockets to 1,500 yards. Then deciding that he no longer could do any better with the skyrocket, he asked for additional funds from the Royal Arsenal to pay for new experiments and the construction of larger and more efficient rockets. As the son of the new Royal Laboratory comptroller, his request was granted. Lord Chatham, master general of his Majesty's Ordnance Board, also ordered the arsenal experts to help Congreve with his research.

Congreve's new plans called for developing a rocket that could be fired accurately at least two or three times the distance of the Indian rocket and that on hitting the target would do more damage than the Mysore rocket barrages.

The first rocket the arsenal developed under his direction was an improvement, but it still fell far short of his goal. He had designed a rocket with a cylindrical body made of sheet metal that was about twelve inches long and three and one-half inches in diameter. To increase the range and speed of the rocket he also developed a faster-burning powder. Congreve found that he had to take extreme care in loading and aging the propellant to prevent it from exploding in the launch tube or during flight. He also learned that aging or storing the propellant increased its performance. He prepared the charge by wetting the powder, carefully jamming it into the bottom end of the cylinder as tighly as possible, and then putting it on the shelf to dry and cure for several months.

His incendiary warhead, which was new also, consisted of saltpeter, sulfur, antimony sulfide, tallow, rosin, and turpentine. This combination was prepared specifically to burn a long time. The mixture was placed in a canvas bag and loaded into the top of the cylindrical body. Hoops were then used to strengthen the thin sheet-metal body. Because these hoops resembled the ribs of a human body, the cylindrical portion was thereafter called a carcass. Another series of wider hoops were soldered to the body in order to attach the long stabilization stick.

To increase the accuracy of his rockets Congreve launched them from a tube. Similar types of tubes are used in many modern rocket weapons, such as antitank bazookas and recoilless rifles.

Congreve conducted numerous tests on these and other rockets of his design at the Woolwich Arsenal range. During these tests he learned how to build a better and longer-range rocket that was less expensive than the artillery in existence. By 1805 he was ready to demonstrate publicly that a six-pound rocket could be shot a distance of about 2,000 yards, or about twice the distance of the Indian rocket.

The term "six-pound rocket" referred to a lead ball weighing six pounds loaded in the body, or cylindrical section, of the rocket. In particular, Congreve wanted to demonstrate his rocket to William Pitt, the king's prime minister. Congreve asked his friend, the Prince of Wales, to help him gain an audience with Pitt. During his visit with the prince at the seaside resort of Brighton Congreve arranged a series of demonstrations, after which the prince, impressed by the

capability of the rockets, decided to support Congreve and
sent him to Pitt. Pitt, in turn, asked Lord Castlereagh, his
adviser, and Lord Mulgrave, an experienced soldier, to witness Congreve's demonstrations.

Pitt and these men were not really impressed by the prince's
support for Congreve's rockets, feeling that the prince was
much too young and inexperienced in military matters to have
any appreciation of military weapons. They were, however,
fair-minded men and decided to see for themselves. When the
demonstration on the Woolwich range proved to be highly
successful, both Mulgrave and Castlereagh recommended to
Pitt that Congreve's rockets be produced by the Royal Arsenal
and used in the war against Napoleon.

At that point Congreve told them he was convinced that a
rocket with a range of 3,000 yards carrying an explosive
charge equal to a ten-inch mortar could be made very light and
portable so that a single soldier could carry the rocket and
launch tube without any difficulties. He also pointed out that
his rockets were much cheaper than mortars and shells and
that they had no recoil when launched. The latter feature
permitted the rocket to be carried and fired from very small
boats as well as large ones, which would add more power to
the British navy.

For the first combat demonstration in 1805, Pitt instructed
Congreve to accompany Sir Sidney Smith in a naval attack
against the French city of Boulogne. This was an obvious
choice, since Napoleon had a fleet of ships and barges poised
at the French port for the invasion of England. But as with
most developments the first use of Congreve's rockets was
less than successful. Ten ships were outfitted with rocket
launchers and sent to Boulogne. There are two accounts of
that mission. The first states that highly unfavorable winds
caused the ships to return without firing a shot; the second
states that the ships fired over 200 rockets setting fire to about
three houses in the town, and that after the bombardment

French soldiers marched around town with the empty car-
casses of Congreve's rockets, making jokes about them.

But very few tests are complete failures. In this instance
the commander of the expedition, Sidney Smith, became a
strong supporter of Congreve after witnessing the rocket
barrage. Smith told his fellow officers that the pointed nose
of the rocket imbedded itself so firmly in whatever it struck
that the warhead was bound to set fire to any wooden ship it
touched. News of his description reached Lord Nelson, then
commanding the British fleet off the Spanish coast. Nelson
asked for Congreve's rockets to use against Cádiz, but unfortunately he never received the weapons that might have
eased England's position in the war.

In winter 1805 war-weary England entered into a very
short armistice with France, resuming its war again in 1806.
At this point England was demoralized by the long hostilities.
Pitt had died. Austria and Russia were defeated at Austerlitz,
and a French invasion force was again being concentrated at
Boulogne for an invasion of England.

To stop the invasion before it began, the second rocket
expedition was sent against that port. This time Congreve's
rockets were true to their mark; they burned the city and
destroyed the invasion fleet. Congreve, watching the rocket
bombardment, said that in about an hour almost 200 rockets
were discharged; the dismay and astonishment of the enemy
were complete. Not a shot was returned, and in less than ten
minutes after the first discharge the town was on fire.

In 1807, during the British siege of Copenhagen, Congreve's rockets proved even more effective. This time the
British fleet, which was lined up off the Danish coast, started
firing at nightfall. Within a short time the sky was lit up as
bright as day, and the greater part of Copenhagen burned to
the ground. That night more than fifteen Danish battleships
and thirty smaller vessels surrendered. It was a crippling blow
to Napoleon, who lost the use of the Danish fleet.

Sir William Congreve; detail from a painting by T. Lonsdale.
(*Royal Artillery Institution, Woolwich, England*)

Congreve, not content just to build the rockets or to stand by as an observer, had directed the launching of hundreds of rockets during the battle. Years later he had a painting made of himself against the background of Copenhagen in flames.

Because of their destructive power Congreve's rockets were regarded by many as horrible infernal machines that should be outlawed in warfare against civilized nations. Lord Wellington, England's great general, stated, "I don't want to set fire to any town, and I don't know any other use for rockets."

There were others who objected to the rocket, not from the humanitarian standpoint but because it was not sufficiently effective as a bombardment weapon. Many British officers complained of the rockets' inaccuracy and the fact that winds sometimes turned the long guidance sticks around, directing the rockets back on the men who fired them. Understandably they objected strongly to being bombarded by their own rockets; however, influential military men and the Prince of Wales continued to support Congreve.

In 1813 Congreve's rockets were sent to Europe and played a major role in the defeat of Napoleon's armies at Danzig and Leipzig. The first use of rockets against Danzig on August 26, 1813, failed to inflict any major damage to the city. The second bombardment in September burned down twenty-three buildings, and the third bombardment in October destroyed all of Danzig's food stores. Following the attacks on Danzig, rockets were used again from October 16 through October 19 during the battle of Leipzig, called the Battle of the Nations.

The key battle was fought outside Leipzig. A force of 325,000 men from England, Russia, Austria, Prussia, and Sweden confronted 214,000 men of Napoleon's Grand Army of France. A British rocket barrage was used to weaken Napoleon's positions for the assaults that followed. After three hard-fought days during which 35,000 Saxons deserted to the Allies, Napoleon ordered a retreat. The army crossed the Elster River over a single bridge. His rear guard was cut off, however, and those who could not swim were drowned or killed in battle. The Allies took 20,000 French prisoners. This battle led to Napoleon's downfall in April 1814 and helped England win a war that it had fought since 1793.

Several months before the final defeat of Napoleon's forces the British formed a special rocket corps. There had been rumors for several years that such a corps would be formed under Congreve's command. But instead Congreve was offered a high position and rank in the regiment of artillery, which he refused, explaining that it was sufficient gratification to have succeeded in developing for British military use a weapon that was his greatest pride.

Along with his refusal of the artillery commission Congreve prepared a twenty-one-page booklet in which he summarized his work and presented in text-book form instructions on the use of his rockets. In the introduction he stated that his rockets existed only because of the support of the prince regent, and since the prince had seen fit to order the formation of a rocket corps, effective on January 1, 1814, he, Congreve, thought it was his responsibility to write the book for the "Instruction of the Officers Corps, for the information of the General Officers of the British Army . . ."

By April 1814, when the long war with France was over, Congreve's rockets were much more destructive than the ones used during the first bombardment of Boulogne. England's wars, however, did not end with the surrender of Napoleon. Two years after his defeat in June 1812, the United States declared war on England because the British were blockading French ports and hurting U.S. trade with France and Europe. The United States also objected to the British use of impressment and search of ships—forcibly removing American and

A painting of the period showing the use of Congreve's rockets launched from a boat. (*The Smithsonian Institution*)

British sailors from American ships and forcing them to serve in the British navy or sending them to prison. About 6,000 such incidents were reported to President James Madison.

In response to these actions and to pressure from other Americans who had visions of conquering Canada easily in order to extend the fur trade President Madison sent a confidential message to Congress on June 1 recommending a declaration of war. On June 18, 1812, Congress approved it, and war with England was under way.

For the next two years there were numerous land and sea battles. In some the United States emerged victorious; in others the British proved superior. Just before the battle for Fort McHenry American forces defeated the British in the land battles of Lundy's Lane, Chippewa, and Fort Erie and in the naval battle of Lake Champlain, where the American vessels *Hornet, Peacock, Enterprise,* and *Boxer* were successful against the British vessels *Chesapeake, Shannon, Arugus,* and *Pelican.*

The war in the south, however, went badly for the United States. On August 19, 1814, the British, after taking control of Chesapeake Bay, landed an army of 4,000 troops on the Maryland peninsula at Benedict. To meet the invasion, headed by British Maj. Gen. Robert Ross, an improvised army of 7,000 Americans under the command of Brig. Gen. William H. Winder took its position at Bladensburg, about six miles north of Washington, D.C. At the last moment the defense plans were disastrously changed by Secretary of State James Monroe. The battle got under way on August 24, 1814, and British rockets were used effectively, frightening the untrained American militia. After two volleys most of the volunteers abandoned the fight, fleeing toward Georgetown. Commodore Joshua Barney, whose sailors and marines had been posted in the battle line, offered the only genuine resistance. His men maintained the position against great odds until Barney was severely wounded and taken prisoner.

The American defeat left Washington exposed to the British. The British forces marched on to Washington and burned the Capitol building and the White House in retaliation for the American burning of the Canadian city of York, later rebuilt and renamed Toronto. The British thought that this would discourage the Americans, but it had the opposite effect and strengthened efforts to defeat the British.

After his easy victories Major General Ross changed his original plan for an immediate attack on Baltimore. Instead he decided to wait for Adm. Alexander Cochrane and then launch a simultaneous attack on Baltimore from land and sea. On September 12 Ross began his land campaign against Baltimore, but in a skirmish near North Point he was killed, and Col. Arthur Brooke succeeded to command. The British fleet under the command of Cochrane cooperated with Brooke in the assault upon Fort McHenry. The fort and 3,500 men under the command of Maj. George Armistead resisted the assault.

After the fort was shelled by cannon for several hours, Cochrane ordered three rocket-equipped ships forward to increase the rocket's effectiveness. Armistead returned the cannon fire, which repulsed the vessels. The British cannon and rocket fire continued until midnight as a diversion for the British troops, but the guns at Fort McHenry halted the British land forces. At seven o'clock next morning, after twenty-five hours of bombardment costing only twenty-eight casualties, the British withdrew. The British had many wounded and killed and had lost a rocket barge.

During the attack Francis Scott Key stood on the deck of a British ship, having been detained there while on a mission to recover a prisoner. In the morning when he saw the American flag still flying, he was inspired to write the immortal lines "the rockets' red glare, the bombs bursting in air," as part of "The Star-Spangled Banner." Although the song immediately became popular, was adopted by the army and the navy as the U.S. national anthem and declared so by a presidential order in 1916, it was not until 1931 that Congress passed a bill actually making it official.

The failure of the United States to seize Canada in several attempts disheartened the "War Hawks," who now wanted peace. England was equally in favor of bringing the war to an end. On December 24, 1814, the Treaty of Ghent was signed, officially ending the war.

Congreve, unlike many other inventors in history, won all sorts of honors for his development of war rockets. When the Prince of Wales was named regent in place of the ailing King George III, he made Congreve his equerry, a member of the royal household, and a lieutenant colonel in the Hannoverian Artillery.

Congreve also had his portrait included in the well-known picture called "Scientific Men of the Day, 1807–1808," was named to membership in the Royal Society, and was elected to Parliament in 1812. After Napoleon's defeat at Leipzig the tsar of Russia decorated him with the Order of St. Anne, and the king of Sweden presented him with the Sword of Sweden.

When King George III died in 1820, the Prince of Wales became King George IV. Congreve was still his senior equerry and spent much time at court. He was reelected to Parliament and held that office for an additional eight years.

He also continued to invent things. As early as 1813 he invented a hydropneumatic canal lock; in 1819 he published his description of a new principle of steam engines. He also developed a gun recoil mounting and a time fuse, built a clock in which time was measured by a ball rolling down an inclined plane, devised a method for enlarging metals, and invented a gas meter; and patented a paper on which unforgeable bank notes might be printed, a process of color printing that became widely popular in Germany, and a method of killing whales with the use of rockets.

Congreve's rockets had an influence on most of the civilized world. Austria, Denmark, Egypt, England, France, Greece, Italy, the Netherlands, Poland, Prussia, Russia, Sardinia, Spain, and Sweden competed in improving rockets during the first half of the 1800s. As far as powder rockets go, Congreve's was as powerful as one could ever be. There remained a problem that scientists of many nations tried to solve, however; that is, to find a better way of controlling the rocket in flight. When the stickless war rocket finally made its appearance, it turned out to be the invention of another Englishman —William Hale, who had solved the problem by placing three metal vanes in the rocket exhaust.

Hale's rockets saw service in the Crimean War (1853–1856) and in the Austrian wars against Hungary, Italy, and Prussia. They were finally introduced into the United States in 1861.

One frightening experience involving Hale's rockets occurred when an agent of Hale's arranged a demonstration for President Abraham Lincoln and some of his cabinet members. The rocket accidentally exploded near the president and his guests, but miraculously no one was injured.

In his final years Congreve's health and fortune took a turn for the worse. He became totally crippled and had to rely completely on a wheelchair. In his later years he became involved in the development of mines in South America and Ireland. Through financial mismanagement he was accused of fraud.

Though declared innocent he was never able to clear his name fully. He left England in disgrace as well as for health reasons and spent the rest of his life in the warm climate of southern France.

Congreve died on May 5, 1828, four days before his fifty-sixth birthday, still believing in a long future for the weapon he had made famous. What he could not know, however, was that the cannon, which was then more than 500 years old and was still a very crude weapon, would shortly undergo many improvements and replace the rocket once again as a weapon of war.

One of the last uses of the war rocket in that century took place in Russia in 1881 during the Turkestan War. These were not the Hale type, with the vanes for guidance, but the stick-stabilized Congreve war rocket.

3

Konstantin Eduardovich Tsiolkovsky

> In my experiments I arrive at many
> new conclusions, but my conclusions
> are greeted with disbelief by other sci-
> entists. These conclusions could be
> confirmed by repeating my experi-
> ments, but no one knows when that
> will happen. It is hard to work alone
> for many years under unfavorable con-
> ditions and receive neither understand-
> ing nor support from anyone.
> Konstantin E. Tsiolkovsky, 1931

Historically, Konstantin Eduardovich Tsiolkovsky followed
in Congreve's footsteps. He was born a year after the Crimean
War ended and was a young man during the Turkestan War
(1881). Both wars involved Russia and the use of Russian-
made Congreve-type rockets.

Congreve spent his working life experimenting with war
rockets—actually building them and testing them. Tsiolkovsky, on the other hand, was academically inclined and as a
mathematician developed some of the earliest scientific computations on rockets as a means of propelling a spacecraft
into outer space.

Konstantin Eduardovich Tsiolkovsky. (*Novosti Press Agency*)

Tsiolkovsky's countrymen consider him the father of Soviet rocketry. His work and his life story have become a legend in his own country. Like most stories about national heroes, after a time it becomes difficult to separate the man from the myth. Fact or fiction, in the Soviet Union Tsiolkovsky's life and legacy are a source of inspiration not only to scientists but also to young people aspiring to careers in science.

Tsiolkovsky's ideas on space travel, or interplanetary communications, were the beginnings of rocket development in the Soviet Union. He is famous for formulating the first concepts of space travel, which inspired future Soviet space scientists as well as scientists throughout the world.

In order to make some of Tsiolkovsky's goals about space

travel practical, however, many complex problems in the field of rocket engineering had to be solved. In fact, Tsiolkovsky's ideas could only be tested after about forty years of technological growth and evaluation and the contributions of several generations of rocket scientists. Only then did Soviet scientists have the technology to test Tsiolkovsky's theories and launch their first earth satellite.

When the first Soviet earth satellite was launched, academician A. A. Blagonravov of the Soviet Academy of Sciences said, "space flight is not the accomplishment of a single man, it involves creative search and the strenuous effort of thousands of people."

Tsiolkovsky was born on September 17, 1857, in the village of Izhevskoye, Spassky Uyezd, Ryazan Gubernia, about 200 kilometers from Moscow. His father, Eduard, was of Polish descent and a forestry expert whose work kept him constantly on the move. He was described as a restless and gloomy person who liked to invent, build, and talk about ideas. As an inventor he once built a fairly large-sized thresher, which failed to work. As a builder he often worked with his sons erecting models of palaces and houses. As a philosopher he readily volunteered his ideas and viewpoints to anyone who would listen.

His mother, Maria Yumashiva, by contrast, was described as a cheerful, witty woman even in the face of many family hardships. She was of Russian descent, with a trace of Tartar stock, and came from a family of artisans of peasant origin. Tsiolkovsky later in life wrote, "My parents were poor, my father, Eduard, an impractical inventor and philosopher. Maria, my mother, as father used to say, had a spark of hidden talent; among her relatives were many gifted people. My father's main trait was force of character and will. My mother's main trait was talent."

Tsiolkovsky was much like any other energetic little boy. He loved to camp out, to climb trees and fences, and to fly kites with boxes containing cockroaches as passengers. He was also known as a daydreamer who imagined there was no gravity. He was an avid reader and a story teller who loved to make up his own stories and at times paid his younger brother to listen to his tales.

But a turning point in Tsiolkovsky's life occurred when he was around the age of nine. At this time he came down with scarlet fever and because of complications became almost totally deaf. As a result, his familiar world of sounds changed overnight. He could no longer play normally with other children and lost his boyhood friends. In fact, "being almost totally deaf," he wrote, "made me a victim of ridicule to the rest of the boys in the neighborhood." At school he could not hear the teachers, so he eventually had to leave school. His deafness kept him from people and prompted him to read, concentrate, and dream to keep from being bored with life.

Because of these circumstances he began a program of self-education. He taught himself from books that were in his father's meager collection and from other books that he could borrow. He mastered mathematics first and then physics.

At the age of thirteen he went through his second major tragedy. His mother died when he needed her understanding and comfort most. After his mother's death his father became gloomier than ever but somehow managed to provide for his large family.

At fourteen Tsiolkovsky began to be interested in boyish inventions and experiments. He built a balloon of tissue paper and tried to fill it with hydrogen to send it up into the atmosphere, but the experiment failed. At that time he was also interested in mechanical flight by means of flapping wings but his model craft never left the ground. Among other things he made a small model of a carriage with sails to be driven by the wind. This was followed by another model of a carriage powered by a steam turbine. Finally he decided to build a carriage large enough to carry him. The project failed, due to the lack of patience and money for supplies.

Then he became very ambitious and decided to build a carriage that he could ride in himself. He spent all the money he could get on lumber, screws, and nails. He discovered that the invention was too complex for him to build himself, but the idea, he knew, was workable.

During this period Tsiolkovsky tells of building a large navigable balloon with a thin metal shell. His interest in this means of flight continued through the years, and at the age of twenty-eight he mathematically studied the problem of dirigibles. Seven years later he finally abandoned the project because of lack of government interest and support.

Tsiolkovsky's father watched his early painstaking experiments with awe. He decided that Konstantin should go on to a higher education, even if it meant sacrifice for the rest of the family. So when Konstantin was sixteen, his father sent him to Moscow with an allowance of about fifteen rubles a month, in the hope that he would manage to enter the technical school there.

Between the ages of sixteen and nineteen Tsiolkovsky lived as a student in Moscow. There he no longer had to put up with the shortage of books that was such a handicap in his home town. Soon after his arrival he began to work regularly in the Chertovsky Library, now the Lenin State Library, one of the largest in Moscow. At the library he became acquainted with N. F. Fedorov, an expert on scientific literature. He admitted to Fedorov that his goal was to read, study, and master the equal of a university course on his own. After that, the library was his school and Fedorov was his unofficial teacher.

He spent the entire first year on elementary mathematics, physics, and chemistry. In the second year he began to study higher mathematics. He covered algebra, differential and integral calculus, analytical geometry, and spherical trigonometry. His motivation was always the same. "The idea of communication with space [space travel] never left me," he wrote. "It was this that stimulated me to study higher mathematics."

During his stay in Moscow he lived in slums and spent a large portion of his allowance on books and on chemicals for his experiments. In his autobiography he wrote, "I remember very well that I had nothing to eat but dark bread and water. I would go to the bakery once in three days and buy nine kopeks' worth of black bread. In this way I spent on bread ninety kopeks a month. For all that, I was happy with my ideas, and my diet of dark bread did not dampen my spirits." His thoughts centered constantly on solving the problems of space flight. One night he thought of a way that a rocket could escape from the earth's atmosphere. The idea of using centrifugal force occurred to him. He likened the situation to a boy whirling a rock on a string around his head. If the rock were released (the rock being like a rocket), it would shoot away from the circle in a straight line. He decided to build the apparatus to test the idea. "I was so excited, even overcome," he wrote later, "that I did not sleep the whole night, but roamed about the streets of Moscow, thinking constantly about the great consequences of my discovery, but by morning I had already convinced myself that it was false. My disappointment was as great as my delight had been. That night left its mark upon my entire life; thirty years later I still dream that I am rising to the stars in my machine, and feel the same delight as on that immemorial night."

So he went back to his books on physics and chemistry. He set up his own problems and experimentally solved them. Acids used in his experiments left spots and holes in his pants, and the boys in the streets could not resist making fun of him. They would shout: "Have the mice eaten your pants?" His appearance was gaunt, and he wore his hair long because he lacked the time and the money to have his hair trimmed.

In the meantime his father had moved the family to Vistka, in northeastern Russia in the Urals. Times were hard, and money was scarce. His father had heard that his son in Moscow was starving himself into a walking skeleton; it was time for Konstantin to return home to his family. In 1876, at nineteen, Konstantin left Moscow for Vistka. The hard life and concentrated reading and studying had affected his eyesight. He now had to wear spectacles. When he returned home, he recalled, "They were glad to see me but astonished at my miserable appearance. It was quite simple. I'd eaten all my fat."

Two years later the family moved back to Ryazan, and Konstantin went with them. His father retired, and Konstantin started to provide for his family by private tutoring. At the same time he continued with his reading and working in his makeshift workshop. Through his tutoring he discovered his ability and desire to talk to children and tell them of natural wonders and man's ability to transform nature. His enthusiasm was such that his father and others thought he would make a good schoolteacher.

In 1879, at the age of twenty-two, despite the lack of a formal education he passed the required examinations and was granted a certificate as a "people's schoolteacher," the lowest rank in the tsarist educational system. The year following his qualifying examinations he became a teacher of arithmetic, geometry, and physics at the Borovskoye district school in the province of Kaluga, some forty-five miles southwest of Moscow.

He married Barbara Sokolova, the daughter of a teacher, in 1881. They had very little money to start a family. His salary as a teacher was meager, and Konstantin needed what he could scrape up to buy supplies for his scientific experiments. His wife, however, used to poverty from childhood, apparently did not mind his using some of the money he made for his laboratory work.

Their circumstances improved when Tsiolkovsky accepted a better teaching position in Kaluga, capital of the province, ninety miles southwest of Moscow. He became a high school teacher in Kaluga and taught there for some forty years.

In addition to his work as an instructor of physics and mathematics in a parochial school for girls, he was given a similar assignment in a boys' classical high school. Neither school gave primary importance to the subjects that Tsiolkovsky was most interested in. Possibly it was just as well. He was left with more time and thought for his own research and inventions in his workshop at home.

At times his students walked in the direction of his house with him after their lessons. He talked to them of wonderful interstellar journeys of the future. "He would say good-bye to us beyond a bridge where, in impossible mud, lay the street at the end of which his house stood," recalled a former pupil, herself an old teacher now. "Rain poured, dusk thickened, but we were reluctant to start back for our homes for this meant no more of that afternoon's inspired, miraculous monologue about mankind's future."

In class, while listening through his antiquated tin ear horn to a pupil's recitation, he would sometimes jump up from his chair shouting, "I've found it." He would throw open his notebook on the desk, hurriedly write down his discoveries, and then, beaming, turn back to the class.

His parochial school girls noticed the awful condition of his winter clothes and shoes and once they took up a collection among themselves and bought their teacher a pair of overshoes. He accepted the gift in silence but was obviously moved by their concern.

During his long career in Kaluga his small laboratory burned completely on one occasion and was flooded on another. He could only rebuild and restock it by denying himself and his family some of their necessities. All of this time he wanted desperately to work on his theories of interplanetary travel. But he was stopped again and again by the many hours he had to give to teaching, by the lack of assistants, and above all by the lack of money. "I was a conscientious teacher and came home from school very tired. Only toward evening could I resume my experiments."

It was in the 1880s when Tsiolkovsky became a teacher in the Borovskoye district school that he began his first scientific research. His studies concerned themselves with three technological areas: the development of an all-metal dirigible, an airplane, and a rocket for interplanetary travel. His approach to all of these problems was as an inventor rather than a builder.

In 1883, ten years after he conceived the idea of conquering space by means of centrifugal force, he came to the conclusion that only by the principle of jet propulsion (rocket power) could space flight be achieved. In his manuscript, "Free Space," he stated that space travel, in the absence of gravity and air resistance, had to be based on the reaction of particles being ejected from a body traveling in space and that motion was impossible without the loss of matter.

In one of his autobiographical articles, Tsiolkovsky wrote, "In 1885 at the age of 28, I resolved to dedicate myself to aeronautics and to develop the theory of a metallic dirigible." During the period from 1885 through 1892 he designed a gas-filled dirigible with expandable sides so that internal gas volume could be increased or decreased according to the ambient air pressure and temperature. In order to expand the gas inside the dirigible it would be run through coils (like a radiator) and heated by passing the engines' exhaust gases over it.

Tsiolkovsky first submitted his idea to the Seventh Aeronautical Section of the Russian Engineering Society, and the plan was rejected on the basis that dirigibles "will eternally be the playthings of the wind." He then submitted his idea to the General Staff of the Russian army, and again the concept was rejected. Before discontinuing his investigations of dirigibles in 1892, he published a paper entitled "Metal Dirigibles." Subsequently, all that he received for his proposals were a few sympathetic reviews.

Although very little recognition was being given to Tsiolkovsky's proposals, his creativity did not stop. During the years 1893 through 1895 he continued to develop his ideas about interplanetary flight in his science-fiction stories *On the Moon* (1893) and *Dreams of Earth and Heaven* (1895). In the latter story he presented his first ideas on the creation of an artificial earth satellite similar to our moon, only very much smaller, and located closer to Earth. He felt that a distance of about 200 miles from the earth's surface was sufficient to maintain the satellite in a suitable orbit.

Tsiolkovsky is also given credit for proposing the idea of an airplane made of metal. In an article written in 1894 entitled "The Airplane, a Birdlike Flying Machine" he described and provided sketches of a monoplane. Tsiolkovsky's plane had wings with rounded edges and a streamlined fuselage. His designs were very much like the structures developed by aircraft engineers about fifteen years later.

He also contributed to research methods by building the first wind tunnel in Russia in 1890. Then in 1900, with a subsidy from the Academy of Sciences, he succeeded in testing some very simple models to determine the force that air puts on a sphere, a flat disc, a cylinder, a cone, and other geometric shapes. Realizing the importance of theoretical wind tunnel experiments, he wrote, "How important it is to formulate the laws of air resistance and friction as precisely as possible. What extensive applications they have to the theory of dirigibles and airplanes." Tsiolkovsky completed more than 1,000 experiments in aerodynamics.

It was during this same period that Tsiolkovsky recognized the value of the internal combustion gasoline engine as a source of power for aircraft. In this regard he wrote, "I have theoretical grounds for believing in the possibility of constructing exceptionally light and at the same time powerful gasoline or oil engines that fully satisfy the requirements of flight."

Tsiolkovsky's airplane, however, like the metal dirigible, was also rejected by official Russian scientific circles. With no support or funds he finally abandoned his work on airplanes.

Not discouraged, Tsiolkovsky turned to rocketry and space travel, for his inventive mind could not stop thinking and researching. In one of his papers he wrote, "For a long time I viewed rockets like everyone else, from the point of view of diversion and minor applications. I don't clearly remember how I first got the idea of performing calculations on rockets; I have the impression that the first seeds, ideas, were planted in my mind by Jules Verne's well-known fantasy, which set my brain to work along now familiar lines. First, desires appeared, and they, in turn, gave rise to mental activity . . . The old sheet with the definitive formulas dealing with jets is dated August 25, 1898, but the preceding makes it evident that I got interested in the theory of rockets earlier, in 1896.

". . . I never claimed to give full solution of the problem. The first inevitable progression of concept, imagination, and story is followed by scientific calculation, and thought is ultimately crowned by accomplishment. My work on space travel belongs to the middle phase of creation. I understand better than anyone else the abyss that separates an idea from its realization since in the course of my life, I have not only thought and calculated, but have also completed projects, working with my hands. It is impossible, however, for there to be no idea: thought must precede execution, and imagination, precise calculation."

As early as 1897 Tsiolkovsky derived his well-known formula that very simply expressed the final velocity of the rocket in terms of the rocket jet velocity and the ratio of the rocket's full weight to its empty weight. At first his formula

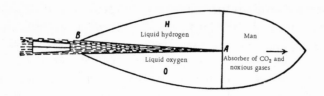

A diagram of the straight rocket nozzle proposed by Tsiolkovsky in 1903. (*Novosti Press Agency*)

gave the rocket's ideal velocity, without taking into consideration flight losses due to drag and gravity. Then later he developed a more rigorous solution that took these two factors into consideration, providing a method for calculating the flight of a missile more realistically.

In 1903 Tsiolkovsky published his classic paper "Investigating Space with Reaction Devices" in the magazine, *Survey of Science.* In the paper he described his formula that made it possible to predict the flight performance of a rocket if the propellants and propellant weights are known, and he proved that it was possible to reach orbital and escape speeds.

His paper was not confined to theoretical calculations; he also gave practical instructions on the designing and building of the individual parts of a rocket, the contours of a nozzle for a jet engine, and so on. He studied the problem of achieving stable rocket flight in a vacuum, examined a great many fuels and oxidizers (substances containing oxygen, which are needed for maintaining combustion in the engine), and made recommendations regarding the best propellant combinations. He also came to the conclusion that single-stage rockets could not achieve the speeds necessary for interplanetary flights. But Tsiolkovsky's ideas were so far ahead of their time that during the first decade of the twentieth century his work went unnoticed. Neither the engineering nor the economic means existed at that time for building long-range rockets.

His theories were treated with indifference and skepticism by his colleagues. Many scientists still considered him an unrealistic dreamer with his head in the clouds and took a very skeptical view of this self-taught scientist without a college degree. Lacking any sort of support, he wrote bitterly, "It is hard to work alone for many years, in unfavorable conditions, without seeing a ray of encouragement or assistance from any quarter."

But Tsiolkovsky, although at times very unhappy, continued working. He tried to identify and solve problems dealing with the design and manufacture of various rocket parts. His investigations included flight control of a rocket in space, the cooling of a rocket engine's combusion chamber, and possible designs for several spaceships.

In 1911 "Investigating Space with Reaction Devices" was printed a second time. This revised reprint of Tsiolkovsky's original article was a popular version of his ideas on space flight and received a wide distribution.

In this article Tsiolkovsky suggested that some new form of energy far surpassing liquid fuel had to be found for space flight. He calculated that rockets powered by liquid fuels could not reach circular orbital velocity, let alone achieve escape speeds for flights to other planets, and proposed the use of atomic power for space flight.

The atomic power he proposed using was not like that being developed today; that is, heat from atomic reaction to heat a propellant, such as hydrogen. He proposed using the energy of atomic decay. He wrote, "It is thought that radium, continually decaying into more elementary matter, emits particles of various masses moving with an inconceivable velocity nearly that of light . . . Therefore if it were possible to speed up sufficiently the decomposition of radium or other radioactive bodies, which probably means all bodies, its use, all other things being equal, could give a jet [rocket] such velocity as to permit it to reach the nearest sun [star] in nineteen to forty years."

During this period Tsiolkovsky also advanced the idea of building an electrojet engine, calling attention to the possibility that with the use of electricity it would be possible to impart great velocities to the particles of the rocket's exhaust.

In 1919, a year after the Russian Socialist Academy of Sciences was organized, Tsiolkovsky was recognized for his work by being nominated and elected to the academy, which was renamed the Communist Academy of Sciences in 1923 and then in 1936 became part of the Soviet Academy of Sciences.

In addition to membership in the academy and a personal pension granted him by the Commission for Improvement of the Lot of Scientists, Tsiolkovsky received help from governmental and social organizations in the publication of his works. In the last seventeen years of his life he published four times as many articles, pamphlets, and books as he did in his first sixty years. Within that period about sixty of his technical and nontechnical articles appeared in print.

During the early 1920s Tsiolkovsky devoted much of his time to the spacecraft reentry problem. In his 1924 article "The Spaceship" he suggested using the denser layers of the atmosphere as a braking medium for vehicles returning from outer space. Tsiolkovsky was not given the credit for being the originator of this concept within the Soviet Union, however. This honor goes to Fridrikh A. Tsander, who expressed it first in his article "Flights to Other Planets," and then somewhat later to Yu. V. Kondratyuk who showed that such a design would result in considerable savings of propellants on reentry.

In his spaceship article Tsiolkovsky also proposed that pressure from sunlight might be used to propel an interplanetary spacecraft. He stated, "But the promise of sunlight, electromagnetic waves, electrons, and helium nuclei (alpha rays) can immediately be adapted in the ether for projectiles that have already overcome the earth's attraction and left the atmosphere and that will later on require an increase in velocity."

He investigated the possibility of using an automobile, steamship, locomotive, airplane, dirigible, gas and electromagnetic cannon, and so forth, as auxiliary means of providing an initial high velocity (lift-off velocity). His studies showed that at best he could obtain only about 300–650 feet per second, which led him to the conclusion "that to give the projectile a velocity of over 200 meters per second [650 feet per second] special means are required. . . . In this case the simplest and cheapest way is to use a rocket or jet; that is, our space vehicle must be placed in or on another land rocket, which, without itself leaving the ground, gives the desired takeoff speed."

In 1924 Tsiolkovsky discussed the subject of multistaged rockets in his book *Cosmic Rocket Trains*. He suggested the use of a two-stage rocket in which the first stage, or the land rocket, "would move on the Earth and in the lower and denser layers of the atmosphere, and the second stage would go on by itself to achieve velocities compatible with interplanetary flight." The lower stage would return to Earth after expending its propellant supply.

During the next two years from 1925 through 1926 he reworked some of his ideas on the construction of nuclear and electrojet engines, but his chief preoccupation seems to have been with finding some way of reaching the high velocities needed for interplanetary flight.

In the last years of his life Tsiolkovsky worked on the development of a theory for jet airplane flight. In his 1930 article "Jet Airplanes" he wrote in detail about the advantages and disadvantages of jet planes in comparison with propeller aircraft. "At an altitude where the medium is 1/100 as dense, its velocity is ten times as great, and it will be twice as advantageous as a conventional airplane." Tsiolkovsky con-

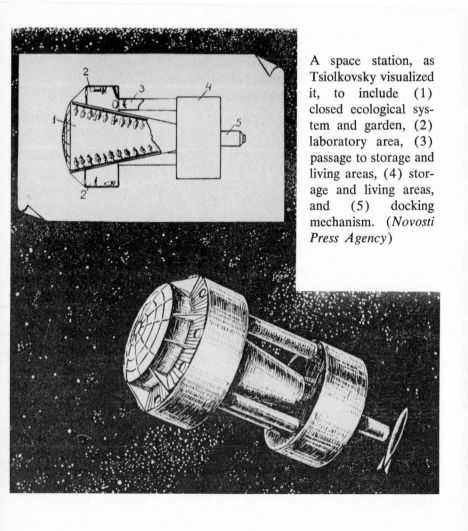

A space station, as Tsiolkovsky visualized it, to include (1) closed ecological system and garden, (2) laboratory area, (3) passage to storage and living areas, (4) storage and living areas, and (5) docking mechanism. (*Novosti Press Agency*)

cluded the article with the statement, "The era of the propeller airplane must be followed by an era of jet or stratosphere airplanes."

In 1935 in another paper, "The Maximum Velocity of a Rocket," Tsiolkovsky once again returned to the problem of multistaged rockets, believing them to be possible and practical. He also realized that technically they were beyond all existing capabilities at that time, stating, "I earlier proposed for this purpose artificial terrestrial roads and rocket trains. Such things are certainly correct and possible but not applicable at the present time because of their costliness and for other reasons. . . . All of these trains and cannons will find use in the distant future, when the significance of interplanetary travel will have grown and will attract more of man's attention, will awaken more faith and real hopes."

He then described a simplified version of a multistage rocket, which he referred to as a "rocket squadron." Unlike the rocket train (one rocket on top of the other), the rocket squadron was a parallel configuration of several rockets connected one next to another.

The squadron design enabled all rocket engines to burn at one time at lift-off. But instead of consuming the entire supply of propellants the rockets would consume only half, at which time the remaining supplies would be transferred into the adjacent rockets, which also would be half empty. The empty rockets would then be jettisoned and return to Earth, and the now full rockets would continue. This process would be repeated until one rocket would finally achieve orbital or interplanetary speeds.

Recognition and honors came late in Tsiolkovsky's life. In fact, in the early days most scientists did not even acknowledge rocketry and space flight as subjects worthy of serious consideration. The problem was compounded by the fact that Tsiolkovsky's articles generally were published in journals that were not read by the established scientists of the day but rather by the engineers and inventors.

Tsiolkovsky himself, later in life, compared his struggle for recognition and acceptance of his work with the struggles of other men of natural science and engineering. In the foreword to an article entitled "Rocket to Outer Space" he noted, "Lamarck wrote a book in which he analyzed and demonstrated the development of creatures from the lowest organism to man. The French Academy with the renowned Cuvier as its head derided the book and publicly called Lamarck a donkey. Galileo was tried, imprisoned, and forced ignominiously to retract his teachings of the earth's rotation. Only by doing so was he saved from the stake. Kepler was imprisoned. Bruno was burned for teaching that there is a multiplicity of worlds. The French Academy rejected Darwin, and the Russian Academy, Mendeleyev. Columbus, after discovering America, was put into chains. The derision of scientists led Mayer to the madhouse. The chemist Lavoisier was put to death . . . There is no counting those that have been burned and hanged for the sake of truth. History is full of such things. And why have academies, scientists, and professionals been condemned to play such a wretched role as extinguishers and even chastisers of truth?"

Despite deafness, poor eyesight, and inadequate funds Tsiolkovsky lived on until 1935, aged seventy-eight, experimenting, writing, and publishing his scientific discoveries.

Just six days before Tsiolkovsky died, he wrote a letter to the Central Committee of the Communist party bequeathing all of his works on aviation, rocketry, and interplanetary travel to the Soviet government. His library manuscripts, models, and other memorabilia were taken to the central offices of Aeroflot but were later put into several museums.

Tsiolkovsky's house in Kaluga is now a museum and contains some of his books and personal belongings. During World War II the Nazis destroyed or took as war booty some of the exhibit material from his home. Most of the exhibits were saved, restored, and replaced after the war, however. Tsiolkovsky's house is now a national tourist attraction to which people from all over the Soviet Union come.

In 1952 a large gold medal in his honor was minted by the Aero-Club de France. Two years later, in 1954, the Soviet government established the Tsiolkovsky Gold Medal, which is to be awarded every third year to the most outstanding contributor to space flight.

In 1957 the Leningrad Studio of Scientific Popular Films made a full-length motion picture about Tsiolkovsky's life entitled *The Road to the Stars*. A monument was erected to

A model of Tsiolkovsky's concept of a rocket ship, from the State Museum of the History of Cosmonautics in Kaluga, USSR. (*Novosti Press Agency*)

him in Kaluga in 1958. It is a silvery upright model of a rocket approximately sixty feet high, which serves as the background for Tsiolkovsky's bronze figure. The granite pedestal at the base of the figure is inscribed "K. E. Tsiolkovsky, 1857–1935," and includes his prophecy from his 1913 paper: "Mankind will not remain on Earth forever, but in its quest of light and space will at first timidly penetrate beyond the confines of the atmosphere, and will later conquer for itself all the space near the Sun."

4

Robert Esnault-Pelterie

When flying became a fact, having
once been only a dream . . . I wondered what the next stage might be:
once the atmosphere had been conquered, there remained nothing more
but to strike out into the empty spaces
of the universe.

Robert Esnault-Pelterie, 1927

Robert Esnault-Pelterie was born in Paris on November 8,
1881. His father was a textile manufacturer. Later in life
Robert Esnault-Pelterie was nicknamed REP, after the initials
in his name. This nickname remained with him throughout
his life.

REP was one of those boys who was interested in mechanical things and very handy with his hands. At the age of
thirteen, when he received a toy steam train as a gift, he con-
verted it into an electrical train by building the entire electrical network himself, including all lighting, switching, and
automatic signals. Shortly thereafter, he assembled his own
small machine shop so that he could build the many things
he needed. In fact, he very seldom bought his equipment but
built what was needed or invented new equipment if no model
existed.

He was an excellent scholar. At twenty-one he obtained a
degree in science (botany, general physics, and general chem-

istry) from the Sorbonne College of Science and Letters of the University of Paris. That same year, 1902, REP was granted his first patent for a highly sensitive electrical relay.

After graduating from the Sorbonne, REP did not hesitate long in selecting a career. Aviation at that time was just beginning in the United States, England, and France; and when news of the Wrights' gliders crossed the Atlantic, REP knew he had found his profession. He would build and fly airplanes.

Esnault-Pelterie was described by his friend André Louis Hirsch, a wealthy banker, as a remarkably good-looking man of great individuality, who even designed his own suits so they could be made with a seam down the front of the trouser legs to keep the creases in.

In describing Esnault-Pelterie's nonscientific interests, Hirsch compared him to American men because of his love of sports and the outdoors and his wonderful sense of humor. According to Hirsch, REP liked to play golf and go horseback riding, and he was interested in camping, automobiles and racing cars. His greatest love, however, was his scientific work.

REP used to rent a small island in the middle of a lake about a mile high in the Pyrenees near the border of Spain. His camp, characteristically, was not the usual camp. The tent floor, which was aluminum covered with rugs, was raised off the ground about two feet; the tent was heated and had electric lights, running water, a bathtub, telephones, and radios.

Naturally the radios were not the usual radios, but ones that were wired so that he could play games with guests when they visited. To this end, a microphone was hidden in the kitchen tent, about one-half mile from the dining tent. When the one o'clock news was presented, he would signal his cook's fifteen-year-old son Pierre to interrupt the regular news broadcast and read some special material he would have prepared for his guests. When the false news item was complete, REP would state, "Wasn't that an unusual news item?" Invariably one of the guests would reply with a knowing air, "Oh, yes, I knew of that already." In recalling these broadcasts he said, "they were very stupid newscasts . . . yet when it was finished, there was always a period of silence from the guests."

In 1928, at the age of forty-seven, REP married Bernaldo de Quiros. Like other incidents in his life, his marriage had an unusual twist. Shortly after arrangements for a conventional home wedding in the presence of family and friends were arranged, REP got an urgent message to come to New York regarding his control-stick patent lawsuit. Canceling their home wedding, REP and his bride were married aboard the *Ile de France* as it steamed toward New York.

Esnault-Pelterie—the pioneer scientist, engineer, and inventor—had one characteristic that was common to all the other rocket pioneers: He liked to work alone. This trait was in no way a handicap for any of them, including REP. His life was very productive in both aviation and rocketry.

REP's involvement in aviation probably began in earnest when Octave Chanute came to lecture at the Aero-Club de France in Paris in April 1903. Chanute, a pioneer in his own right, told of his gliding experiments and described the Wrights' glider in great detail, emphasizing the awkwardness of the control system, which required operating the rudder and wing-warping controls at the same time.

Two of the men in the audience were Esnault-Pelterie and Ernest Archdeacon, a wealthy automobile and balloon enthusiast. These men had already, in 1903, supported another engineer, Léon Levavasseur, in his attempt to build an aircraft with an allotment of less than $400 from the French Ministry of War. On completion, the aircraft resembled a poised pterodactyl. It had two propellers, one in front of the other in back of the cockpit, turning in opposite directions; it was mounted on skids, with two small wheels designed to run along launching rails. The airplane, however, could not fly. Levavasseur, embarrassed by his failure, removed the motor, dismantled the airplane, and burned the remains.

When he later recalled this period of his life in aviation, REP said, "Having started with aviation as early as 1903, in an attempt to check the Wright brothers' results [REP made some of France's first experimental flights in a double-wing glider], I promptly abandoned their biplane type to devise the first monoplane, bearing in front a seven radial cylinder engine; at the rear were two rudders and a fin."

His first attempt at building a flying machine failed. In 1904, to test the Wright design, REP and Archdeacon each attempted to build a Wright type of glider. Both men failed in this because they may have lacked adequate construction details. But REP's interest and enthusiasm never diminished, and he went on to the next project. Having come to the conclusion that wing warping was impractical and dangerous, he began experiments that led to his invention of the aileron. The first step was fitting two independent horizontal rudders to the wing tips of the second Wright glider he constructed in 1904. This glider also failed because of other shortcomings.

Undaunted, he kept building and experimenting. In 1904 he designed and constructed a tailless biplane glider of his own design. He tested this design by towing it behind an automobile at speeds of sixty miles per hour. Still other sections of the airplane were tested by mounting them above the auto.

In 1906 REP completed his first successful monoplane, and his inventive genius was evident everywhere in the airplane's design. In contrast to the Wright Brothers' biplanes—which were constructed of wood and fabric and powered by two engines with wooden propellers that pushed the plane, REP's airplane was all metal in design. It had a steel-tube frame and streamlined fuselage; it was a monoplane powered by REP's specially designed air-cooled radial engine with a metallic propeller that pulled the airplane.

REP knew that his monoplane had to be constructed from very light materials of great strength. This particularly applied to the engine. So he built his own—an air-cooled engine with seven cylinders mounted in a circular fashion, rather than one cylinder behind the other as was conventional. He thought that by this arrangement he could provide a very lightweight engine with a reliable ignition system, and he was right. Thirty years later almost all of the aircraft built in the world used engines based on REP's 1906 design.

Another inventor might have stopped at this point in the design, but REP did not. He went on to invent the metallic propeller for his monoplane. His report on his metallic propeller theory, presented to the French Civil Engineering Society in 1906, won him the society's annual grand prize—a gold medal.

REP's airplane design had other original features that were in later years to become standard equipment in most aircraft. One was the control stick (a single "levered rod" that controls the elevators essential in banking an airplane during a turn or maneuver); another was the two-wheel axleless landing gear; and still another was the hydraulic brake.

REP's first successful plane, which was the model for all planes built during the next fifty years, still can be seen. It is the only airplane the French government has kept on permanent display in the Musée Conservatoire National des Arts et Métiers in Paris.

The year 1906 was especially important to REP, not only for the success of his first airplane but also for being granted a pilot's license, the fourth one issued by the Aero-Club de France. A year later he piloted his own plane in test flights.

REP constructed one of the first aircraft factories in France that was specifically equipped to mass produce aircraft. From 1908 through 1914 a number of his planes took

The Esnault-Pelterie 7-cylinder, 90-horsepower airplane engine, manufactured in 1907. (*The Smithsonian Institution*)

part successfully in races and competitions, setting many new records. He liked to enter his planes in these races, which were very popular at the time, since, much like car races today, they were a proving ground for new ideas.

REP was awarded patents for many of his ideas. Examination of some of these patents shows that he was especially aware of pilot safety. (He survived a serious crash in 1908.) He was granted patents on safety belts, speed indicators, parachutes with quick-release mechanisms, and double controls for pilot instruction in air schools. He was one of the first to test the individual parts of his aircraft for structural strength before he put them together. In essence, he created most of the elements of modern aircraft.

In 1909 *The Daily Mail* of London offered a prize of £1,000 (about $5,000) to the first aviator to fly the English Channel. Many tried and failed, but a Frenchman, Louis Blériot, finally won the prize. Blériot was an energetic hawk-nosed manufacturer of automobile headlights who occupied himself passionately with aviation.

On July 25, 1909, Blériot took off from France in a plane with no compass or navigating instruments of any kind. He flew hunched over his controls with the airstream from the propeller whipping past his goggled face. High winds over the Channel forced him off course, but he landed near Dover Castle, in almost the exact location that Jean Pierre Blanchard and John Jeffries had started their successful cross-Channel balloon flight in 1785.

France's airplanes were the best in the world in those years. Their superiority was attributed to such French aircraft builders as Henri Forman, Levavasseur, and Esnault-Pelterie, among others.

REP continued to build aircraft until 1914. With the future of aviation very well established, however, REP's thoughts raced ahead, visualizing the next step as the natural extension of aviation know-how to the conquest of space.

Robert Esnault-Pelterie at the controls of the REP-2 monoplane. (*The Smithsonian Institution*)

Lise Blosset, the director of the Information and Documentation Division of the Centre National d'Études Spatiales, or CNES (the French equivalent of our National Aeronautics and Space Administration, or NASA), states that this interest in space began about 1908, which is confirmed by published accounts of REP's studies in this field.

Many years later, in a lecture before the French Astronomical Society on June 8, 1927, Esnault-Pelterie recalled his observations, dating back almost two decades, about the future of aviation and space. In this lecture, which appeared in print in an expanded form in 1930 under the title *L'Astronautique,* REP stated, "When flying became a fact, having once been only a dream, it was apparent to me, as one who remembered the time when there were even no automobiles, that it would develop rapidly, and I wondered what the next stage might be: once the atmosphere had been conquered, there remained nothing more but to strike out into the empty spaces of the universe . . . Would this be possible? . . . The answer to this question lies in the laws of momentum, but let me add that due to the enormous initial mass needed, I regard the trip to the moon, at least at present, no more than a theoretical possibility."

In February 1912 Esnault-Pelterie gave his first known scientific lecture on related space travel topics in Saint Petersburg (now Leningrad). He repeated a reading of the paper before the very highly regarded French Physics Society. It was the beginning of his efforts to introduce the idea of rocketry and space travel into the world of science.

In his lecture he once again tried to destroy the misconception that a rocket cannot operate in a vacuum. He discussed the theory and importance of mass ratios and in the face of skepticism among French physicists showed that it was theoretically possible for a specially designed spacecraft to travel from Earth to the moon.

His paper also presented the concept of using auxiliary propulsion for the control and stabilization of spacecraft, as well as the incremental speeds required by a rocket in its various phases for round-trip voyages to the moon. He had made calculations of the durations of trips to the moon, Mars, and Venus and discussed the heating problem of spacecraft—being heated by the sun on one side and being frozen by the lack of heat on the other side. He concluded with a prediction that interstellar space travel would be made possible once atomic energy was mastered.

During a lecture in 1912 Esnault-Pelterie mentioned the work of Dr. Robert Goddard at Princeton and Clark universities and his calculations on methods of reaching extreme altitudes. Unknown to REP at that time was the work of Dr. André Bing, who was awarded a Belgian patent for apparatus for exploring upper atmospheric regions.

By the early twentieth century rocketry was an idea whose time had come. Thus it is not surprising that five major rocket pioneers were exploring the new frontier of space independently yet covering somewhat similar ground in their research. So it is with all science and technology, even today.

In 1912 REP's lecture was published in the *French Physics Society* journal, but because of the page limitation and editing by the journal's secretary, the article was not very accurate or readable. The edited version was responsible for creating an apparent difference between Goddard's and REP's calculations concerning the possibility of building vehicles capable of escaping from the earth's gravitational field.

In 1915 a Russian author by the name of Aleksander B. Shershevsky, writing in German, claimed that Esnault-Pelterie and Tsiolkovsky held a debate on space travel for the tsar of Russia, Nicholas II. Willy Ley, the late international space writer, trying to confirm whether or not this debate had really taken place, wrote to both men.

Esnault-Pelterie replied immediately, saying that the rumor might have started as a result of his 1912 lecture in St. Petersburg. He assured Ley that he had never seen the tsar, nor had he ever met Tsiolkovsky. Tsiolkovsky's letter, which arrived somewhat later, confirmed REP's letter.

Following World War I REP sued France, England, Germany, and the United States for infringing on his control-stick patent. He won damages from all he sued, and the corporations that had made World War I airplanes had to pay REP royalties. These royalties allowed him to return all of his father's money, which had been invested very heavily in his airplanes. Costly experimentation and taxes depleted the remainder of his fortune.

In 1920 Esnault-Pelterie resumed the theoretical work on the flight of rockets that he had put aside during the war. On June 8, 1927, he gave a lecture at the Sorbonne on the progress of his research. The lecture, enitled "Exploration of the Very High Atmosphere and the Possibility of Interplanetary Travel," updated his earlier work dealing with velocity requirements of a rocket for many different types of missions and the importance of the ratio of the rocket's initial mass to its final mass. The lecture also included detailed information on the flow of gases through a rocket nozzle.

In seeking to name this new science dealing with space travel Esnault-Pelterie tried to find an appropriate Latin prefix. The word "aviation" had been derived from the Latin prefix *avi*, meaning "bird." The Latin for "star" is *sid*. So from this REP created the word *sideration* to mean "space science." The word apparently was too obscure, however, and was never accepted.

Another Frenchman, J. H. Rosny, Sr., a novelist and science-fiction writer, suggested "astronautics." The word was first used in December 1927 at a meeting of the French Astronomical Society. At this meeting it was suggested that a committee be created to gather all available documents and study what was happening in the field of rocketry. This committee was designated the Astronautics Committee, and the word subsequently was accepted internationally.

On February 1, 1928, Esnault-Pelterie, then one of the most distinguished figures in rocketry, and the wealthy French banker, André Hirsch, joined forces and founded the highly regarded REP–Hirsch International Astronautics Prize. The sum of 5,000 francs (about $200) was to be awarded yearly for the best original scientific work, either theoretical or experimental, that would advance the state-of-the-art in space travel or in any of the branches of science that are included in astronautics.

The award committee included some of France's outstanding scientists. One of the members was the writer Rosny, then president of the Academy Goncourt, a nationally famous literary group of ten members that awarded an annual prize to the French author who produced the best work of fiction during the year. The award was to be announced and pre-

sented by the French Astronomical Society—the first scientific
society in the world to recognize astronautics as a new science
with a future.

Rocket pioneers all over the world were pleased to hear
about the REP–Hirsch prize, particularly because of Esnault-
Pelterie's international reputation and the influence he would
have on others in astronautics.

The first prize was awarded to Hermann Oberth, the
German rocket pioneer, for the publication of the second
edition of his first book. To emphasize Oberth's outstanding
contribution the prize was doubled that year.

No prize was awarded in 1929 or 1930. In 1931 it went to
the French engineer Pierre Montagne; it was not awarded in
1932; and in 1933 Montagne again was the recipient. In
1934 the French engineer Louis Damblanc received a "prize
of encouragement," and in 1935 the award went to a Russian
named Ary Sternfeld, who wrote to Hirsch after the Sputnik
launch in 1957 that key sections of REP's books on space
travel, translated by Nikolai Rynin, had played an important
part in the Soviet space program and that they had used his
mathematical theories in their work. Alfred African, president
of the American Rocket Society, received the 1936 award
jointly with the society. The prize was for the experimental
work of the ARS and for African's design of a high-altitude
rocket.

REP worked from 1926 to 1930 on *L'Astronautique*. It is
considered to be the basic text on astronautics and is almost
an encyclopedia in the diversity of topics it covers, ranging
from rocket motion, density and composition of the upper
atmosphere, and flow of gases through a nozzle to the possible
use of rockets—including high-altitude exploration, launching
probes to the moon, high-speed travel around the earth, and
entry through the atmosphere—and interplanetary travel.

The book recommends using a 100 percent oxygen atmosphere for the spacecraft, which would reduce the internal

pressure to about one-tenth atmosphere and, in turn, reduce leakage. He also mentions the use of gyrostabilization of the spacecraft and retro-braking on reentry, as well as the use of parachutes to lower the spacecraft to earth.

In a speech made on May 7, 1930, at a luncheon sponsored by the French Astronomical Society, REP remarked that in preparing *L'Astronautique* he had concluded mathematically that a trip to the moon could be made within fifteen years— a fairly accurate prediction considering the interruption caused by World War II.

In that same speech REP confirmed his agreement with the German scientist Hermann Oberth that using liquid oxygen and hydrogen as propellants one could get jet velocities greater than 13,000 feet per second. (These propellants were used years later to launch Apollo.) He went on to say that "It would first be necessary to construct rockets supplied with instruments which would be sent empty to a height of 50,000 to 70,000 meters [more than 36 miles]. Afterward ascensions would be made with a crew aboard. At the start the pilot would fly almost parallel to the earth . . . The distance of these attempts would be gradually lengthened and in fifteen years' time I feel confident that it would be entirely possible to make a voyage around the moon and return."

Late in 1931, following an extended trip to the United States, REP was experimenting with the development of a liquid rocket engine of his own design. During one of his tests using tetranitromethane as a propellant the engine exploded and REP lost four fingers from his left hand.

After the accident the French Air Ministry provided REP with a subsidy to continue his work. The support was limited, however, with only enough money to study a few mechanical devices but no funds to build them. In addition, his work was now classified by the military. This was extremely unfortunate because a few years later, in 1940, when the Germans marched into Paris, the records of his work were either

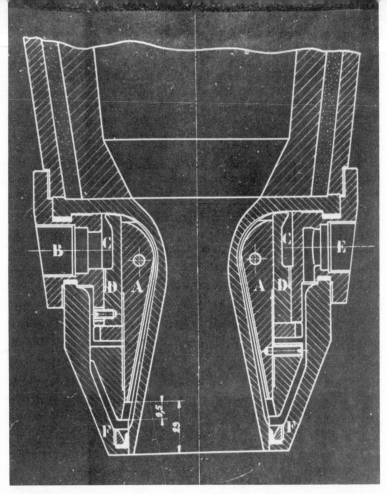

A diagram of the rocket nozzle designed and tested by Esnault-Pelterie in 1936. (*The Smithsonian Institution*)

lost or destroyed in order to prevent them from getting into the hands of the enemy.

After the accident with the tetranitromethane REP returned to liquid oxygen, which he had earlier considered particularly dangerous to handle. His studies quickly established the proper mixture ratio of liquid oxygen and hydrocarbon fuel to get the best performance in the engine.

Working in his laboratory in 1932, his next goal was to

develop an engine suitable for the propellants he had just tested and to build a test stand for the engine. He completed both, and from 1934 through 1937 he experimented with the engine and the propellants. In 1937 an official test was arranged for a number of French dignitaries. The engine operated with a thrust of 275 pounds for sixty seconds. The engine itself was qualified, but REP was refused any additional subsidies. He offered to construct a gyroscopic stabilization device essential to the flight of a conventional rocket and a finned rocket that would not need gyroscopic controls. Both were rejected, and REP terminated his project. The outbreak of World War II in 1939 put an end to his activities in astronautics, and when the war ended, REP, who had retired to Switzerland unknown and misunderstood, abandoned space research. The loss and destruction of almost all his unpublished works were a great loss to astronautics.

However, on May 9, 1947, REP gave his last lecture at the Aero-Club de France on the results of his propellant studies. To the older list he added liquid hydrogen and liquid oxygen, uranium and plutonium.

His last years were very difficult for him. He was harassed by the tax department, and at one point his furniture was attached. After a very rich, fruitful, and exciting career that spanned the pioneering period of both aviation and space travel he did live to see his ideas reach development, first in the German's V-2 rocket and later in the Soviet's launching of the first Sputnik. On the day of his death, December 6, 1957, a Vanguard rocket was launched from Cape Canaveral, like a salute in his honor.

Lise Blosset of the CNES, who has prepared an extensive technical bibliography on Robert Esnault-Pelterie, had this to say: "After looking into the laborious life of this ingenious pioneer, we salute the memory of this universal man, hardly knowing what to praise the most, the researcher's rich imagination, the theoretician's rigorous reasoning, the experimentalist's capability, boldness, and intrepidity, or the engineer's concern for perfection."

André Hirsch, who knew REP best and encouraged him throughout his working career, remembers him as "a man who read and thought only of science . . . pure science. . . . I was his only friend. He was a very lonesome sort of a man."

5

Robert Hutchings Goddard

How many more years I shall be able
to work on the problem, I do not
know; I hope, as long as I live. There
can be no thought of finishing, for
"aiming at the stars," both literally
and figuratively, is a problem to occupy
generations, so that no matter how
much progress one makes, there is
always the thrill of just beginning.
                    Robert H. Goddard, 1932

Dr. Robert Hutchings Goddard was born in Worcester,
Massachusetts, on October 5, 1882. He came from an old
New England family and in American Yankee tradition was
full of determination and handy with mechanical things. He
was a careful scientist who reported his findings only after
thorough systematic research, and then not to newspapers
for public acclaim but to qualified scientific organizations,
among them The Smithsonian Institution in Washington, D.C.

Casual visitors to his laboratory saw him as a man who
avoided the attention of others by keeping himself in the
background and minimizing his own accomplishments. But
his wife and friends saw another Goddard. To those who knew
him well he was warm and friendly, a man of strong will,
with an interest in art that displayed itself in his ability at
the piano and his oil and watercolor paintings.

As a boy, Goddard showed early interest in science and experiments. He enjoyed playing with kites and watching the flight of birds. His father gave him a telescope, which he used for years to observe distant hills, the moon, and the stars. Since he lived in the country, he could carry a slingshot much of the time, and later he had a rifle, so that he became an excellent marksman.

Like many science-minded boys, he played with chemicals and had the usual mishaps with them. Once when he was trying to make diamonds by chemical means, he caused an explosion that frightened the family; another time he filled the house with billowing smoke.

In fall 1899, when he was seventeen years old, he climbed a cherry tree behind the house to trim some branches. As he worked, he began thinking about how to make some device that had the possibility of traveling to Mars and wondered how it would look on a small scale, if sent up from the meadow below. He wrote later that he was a different boy when he descended from that tree, for now he had a very definite purpose in life.

In high school he found mathematics difficult, but he knew that if he wanted to invent something that would go higher than anything had gone before, he had to know algebra, geometry, and much more. He led his class when he graduated, and in his graduation oration he closed with the much-quoted line, "It has often proved true that the dream of yesterday is the hope of today and the reality of tomorrow."

In 1904 he entered Worcester Polytechnic Institute to study for a career in physics. In his first years as a college student Goddard demonstrated his vision and imagination in a paper entitled, "Traveling in 1950." His inventive mind conceived of a railless tube train supported by only a magnetic field. He boldly proposed that people could ride on this type of train from Boston to New York in ten minutes.

By the time he was a senior in college Goddard was convinced that rockets would be better than balloons for reaching higher altitudes, and during that year he began his first experiments with rockets. Whenever he could spare time from his studies, Goddard could be found in the science laboratory, bench testing the efficiency of the powder used in the then known powder rockets. During these tests the powder often gave off clouds of choking black smoke, quickly filling the small basement laboratory. Unperturbed, he patiently continued his experiments, making precise measurements of the energy contained in powder rockets.

Although Goddard's experiments were confined to the only rockets available in his day, his vision extended far into the future. In the same year that he began his experiments he wrote an article suggesting that the heat from radioactive materials might be used in interplanetary spaceships.

After graduating from Worcester Polytechnic with a bachelor of science degree, he was asked to remain as an instructor in physics. While teaching, he enrolled at Clark University, also in Worcester, obtaining his master's degree in 1910 and his doctorate in 1911.

As a graduate student at Clark, he came under the influence of a great professor, Dr. Arthur Gordon Webster, who was trained in Berlin under Hermann von Helmholtz. Webster gave him the background he needed for advanced theoretical and experimental work in physics. As a result of his influence Goddard continued to teach physics, first as an instructor and later as a professor, when he was not engaged in rocket research.

As a young physics instructor at Worcester Polytechnic, Goddard made some important contributions in the field of electronics. In recognition of this outstanding work he received a one-year research fellowship to Princeton University in 1912.

At Princeton Goddard made some of the theoretical rocket computations that were later to lay the foundation for his lifelong work in rocketry. Near the end of his year's fellowship, however, overwork brought on a severe physical breakdown, and he contracted tuberculosis. The doctors gave him two weeks to live. His nurse reported that she found bits of paper with figures on them under his pillow, which he refused to give up, saying, "I have to live to do this work." Contrary to the doctors' expectations, Goddard's health began to improve.

After more than a year of recuperation he returned to Clark as a part-time teacher. Offers came from Princeton and Columbia universities, which he regretfully refused, feeling that the teaching load in such large schools would reduce the time available for his research and experimentation.

Goddard had measured the performance of powder rocket motors at Worcester Polytechnic; at Princeton he made his theoretical flight calculations; and at Clark, and Mt. Wilson near Pasadena, California, he experimented with naval signal rockets. With this experience he began to build rocket motors of his own design. He abandoned the use of black powder and was probably the first person to use smokeless powder as a fuel for a rocket motor. He then changed the rocket exhaust nozzle to a smooth, tapered cone shape. He applied for his first two patents: one for a cartridge-loading rocket and the other for a two-stage powder rocket.

From 1914 through 1916 Goddard continued to make improvements in the design of a rocket motor. He had worked alone in a small workshop with limited funds, spending all the money he could spare from his small salary of $2,000 a year to buy materials and equipment. But by 1916 Goddard had reached the limit of what he could do with the meager sum of money available to him. He needed more equipment and larger facilities.

He could not stop now. His experiments, calculations, and improvements on the design of rockets were technically sound. His vision of what the future held for rocketry was real and urgent to him. Although he was working toward the development of rockets to explore the upper atmosphere, he also knew that this was only a step for man before he ventured into outer space, to the moon and beyond.

He was now in a position to apply to various foundations for financial backing. His paper presented the mathematical equations he had worked out at Princeton in 1912 and 1913 and described his experiments with solid propellants and the test results he had obtained. In the latter part of the report he discussed such advanced ideas as rocket staging and the recovery of instruments from altitudes. He sent these facts, neatly bound, to two likely research foundations. His report brought forth no reply from the Aero Club of America but the Smithsonian Institution was interested and asked that the manuscript be sent to it.

Fortunately Goddard was not easily discouraged, and on September 27, 1916, he wrote a six-and-one-half-page letter to The Smithsonian Institution in which he stated, "For a number of years I have been at work upon a method of raising recording apparatus to altitudes exceeding the limit for sounding balloons; and I feel that I have settled every point upon which there could be reasonable doubt. Incidentally, I have reached the limit of the work I can do single-handed, both because of the expense, and also because further work will require more than one man's time." He went on to mention the military potential of his device as a long-range weapon (Europe was engaged in World War I) but expressed the belief that "exclusive use of the device for warfare would be a loss to science."

In closing he asked if the Smithsonian might have his proposed method and techniques reviewed by a scientific committee and, if favorably received, funds found to support further research.

When Goddard's letter arrived on September 29, the secretary of the Smithsonian, Charles D. Walcott, was on a business trip, and the letter was brought to Charles G. Abbot, the acting secretary. Abbot was immediately intrigued. He wrote a note to Walcott, directing attention to Goddard's specific requests.

On October 11, 1916, Walcott answered Goddard's letter, indicating his interest and inquiring about the amount of funds Goddard was seeking. Goddard responded with a year's budget estimated at $5,000. More details were requested by Abbot, and Goddard sent copies of his patents and a lengthy manuscript and offered to come to Washington to brief them on his work. On December 18 Abbot wrote to Walcott recommending that Goddard's work be sponsored.

An independent assessment of Goddard's work was requested from the Bureau of Standards in Washington. Dr. Edgar Buckingham, a theoretical physicist at the bureau, agreed with Abbot and expressed the "hope that The Smithsonian Institution will see fit to help Mr. Goddard in developing his invention."

On the basis of the two favorable opinions Walcott wrote to Goddard on January 5, 1917, that The Smithsonian Institution had verified the soundness of his work and the accuracy of the numerical data, that they were favorably impressed with the ingenuity of his mechanical and experimental capability, and that a grant of $5,000 from the Hodgkins Fund was approved. Reports were to made "yearly or oftener if notable progress" was made, and a part-payment check for $1,000 was enclosed.

This letter was the start of a long friendship between Charles G. Abbot of The Smithsonian Institution and the physics professor Robert H. Goddard.

When the United States entered World War I in 1917, Goddard pointed out to the Smithsonian the possible value of his rocket concept to long-range bombardment. Abbot recommended to the War Department that it spend a sum not exceeding $50,000 under his direction for experiments. Goddard wrote, "if the apparatus has any possibilities as regards warfare . . . it should be ready for the drive by the Allies, which will probably take place next spring."

On January 22, 1918, the secretary of The Smithsonian Institution and the director of the U.S. Bureau of Standards jointly signed a letter to the chief signal officer, U.S. Army, enclosing a report on Goddard's research and requesting the sum of $10,000 for development. Since Walcott was chairman of the National Research Council, in addition to his Smithsonian position, this request was accepted, and Signal Corps support was assured.

In the first ten months Goddard was a very busy man. He employed seven men, equipped a shop and laboratory at Clark, struggled to obtain special gun steels, and drew up rocket designs. He also performed tests, collated data, and wrote reports.

Abbot was also kept busy. When the Worcester draft board was about to draft a key workman into the army, Goddard appealed to the Smithsonian for help. Abbot, after much effort, managed to get the man's draft classification changed. When Goddard needed special test gauges, Abbot obtained them. Still other problems arose after the test work was relocated to Pasadena, California, in June 1918. A new shop had to be equipped and staffed, and special materials and powder formulations had to be obtained. The Smithsonian and Abbot provided whatever was needed, including hospitalization, medical payments, and a satisfactory settlement when one of Goddard's young assistants, Clarence N. Hickman, was badly injured in an explosion accident.

By handling most of Goddard's administrative needs, Abbot permitted Goddard to concentrate on his scientific work. As a result, within a short time Goddard was able to develop two rockets successfully: one in which propellant charges were injected into a combustion chamber, similar to a machine gun, and one that launched high-speed rockets from a launching tube.

During a demonstration at the Aberdeen Proving Grounds on November 6 and 7, 1918, Goddard completely amazed the military viewers with his tube-launched rocket, which was capable of going through a stack of sandbags. But World War I ended several days later, and the bazooka-type rocket was not used. (Twenty years later, during World War II, a similar type of weapon was used as a very effective antitank weapon.)

At the end of World War I, Goddard went back to Clark to teach physics. Attempting to wind up his rocket research activities, Goddard wrote to the Smithsonian on April 8, 1919, suggesting that they publish a report of his concept on high-altitude rockets. The Smithsonian published his paper under the unassuming title of "A Method of Reaching Extreme Altitudes."

This paper, in which Goddard recorded his calculations and experiments indicating the possibility of reaching the moon, was greeted with an avalanche of ridicule, skepticism, and abuse, especially in the Sunday supplements of newspapers. Goddard, who was sure of his facts, replied to many of his nonscientific critics with polite scientific statements. After all this unfortunate publicity, however, Goddard throughout the rest of his life was very careful about disclosing the results of his work and sought rather secluded surroundings for his research and experiments.

Despite the unfair criticism Goddard rallied and became even more engrossed in his work. In fact, Clark students told a story about his walking along a corridor of the university on a rainy day still holding an open umbrella over his head. He never used any illness as an excuse for idleness. It was only the social and recreational side of his life that was curtailed.

After his return to Clark Goddard changed his technical approach and switched from solid fuel to liquid propellants. He knew that if a rocket was to go faster and travel higher, better rocket propellants were needed. Thus, he began experimenting with liquid oxygen and various hydrocarbon fuels, such as kerosene and gasoline. Learning to use and store oxygen at $-300°$ F in light tanks presented an enormous challenge.

With small grants from Clark University (1920–1922) and then from The Smithsonian Institution (1922–1930) Goddard was able to overcome the difficulties and build for the first time in history a workable liquid rocket engine. It had valves and tubes that allowed two propellants to be fed separately under pressure and mixed when needed in the rocket combustion chamber just before ignition.

On November 1, 1923, Goddard tested the first liquid rocket in a specially built stand, using liquid oxygen and gasoline; these same propellants and the same method of pumping propellants are used on many of our modern rockets.

During summer 1924 Goddard married Esther Christine Kisk, whom he had met when she was secretary to the president of Clark University. She was an untiring helper in his work, and she kept notes and made photographic records of his experiments throughout the rest of his career.

By December 1925 Goddard test fired a liquid rocket engine that used gas pressure rather than pumps for feeding propellants. The gas pressure method of supplying fuels was a simplification of the pump method and is identical to that used in the second stage of the Vanguard and other rockets.

Then on March 16, 1926, an unheralded but history-making event took place. The first liquid propellant rocket in the world was launched. Goddard, bundled up on that cold windy day out on his Aunt Effie's farm, fired a liquid propellant rocket that flew a distance of 184 feet in 2.5 seconds.

Robert Hutchings Goddard, in a photograph taken March 16, 1926, shown with the first liquid-propellant rocket. (*The Smithsonian Institution*)

This flight, as unspectacular as it may seem, is considered a benchmark in flight history as great as that of Orville Wright, who in his first flight achieved a distance of only 120 feet.

Unlike our present rockets, this famous little rocket had its engines above the propellant tanks and the propellant tanks connected to the rocket engines by large feed pipes. At takeoff it weighed almost 10.5 pounds. Of this amount, the structure was 6.0 pounds; the oxygen, 3.7 pounds; and the gasoline, about 0.75 pounds. The thrust of the motor was about 9.0 pounds. On firing, the rocket remained on the launch stand for some seconds, and as the weight of the rocket became less than the engine's thrust, it lifted slowly on its short historic journey. Unfortunately Esther Goddard was recording this event with a motion picture camera that held only seven seconds of film; thus, the film had run through before takeoff.

After collecting and studying all of the pieces after the flight, Goddard reduced the length of the engine nozzle, increased the engine throat diameter, added some additional braces to the structure, and flew the rocket again on April 3. Two more attempts were made on April 13 and 22, but eventually the hot propellant gases burned a hole through the walls of the combustion chamber.

By May 4 the rocket parts were rearranged. The engine was relocated to its more classical position at the rear of the rocket, eliminating the need for long propellant feed lines, which had added weight to the earlier design. It was this later design that Goddard gave to the Smithsonian, and it is on display today, together with a copy of his first liquid rocket (March 16, 1926) and his larger rockets.

Goddard quietly reported the results of this historic rocket firing to The Smithsonian Institution two months later. To appreciate the importance of this flight one must remember that today's satellite space stations and moon rockets use similar rocket propellants, liquid oxygen and kerosene, that Goddard first experimented with in his 1926 rocket.

In the next three years, from 1926 to 1929, Goddard patiently tested and improved his rocket designs in terms of the engine's performance and rocket weight reduction. In 1927 he built a rocket weighing about 150 pounds and producing 200 pounds of thrust, and in 1928 he built a smaller, simpler rocket with parts that were easy to replace when they wore out. The smaller rocket had a thrust of about forty pounds and a length of about eleven feet.

Two flights were made with this rocket, which was launched

at nearby Auburn, Massachusetts. One on December 26, 1928, and the other on July 17, 1929. The July 17 flight carried a small payload consisting of a thermometer and barometer together with a camera to record data at the peak altitude of the rocket's flight. Leaving the tower, the rocket rose skyward with a tremendous roar. Unlike his unpublicized 1926 rocket firing, this rocket caused a great deal of excitement. The local people heard the blast and thought a flaming airplane had descended upon them. The rocket and the commotion it caused was reported in *The New York Times*.

In the interest of public safety and to avoid any more unfavorable public reaction Goddard decided to find a new testing site. For a short time the army allowed him to use their Fort Devens, Massachusetts, artillery range, but he could only fire after a rain or if there was snow because authorities were concerned about forest fires. Once again the Smithsonian paved the way with letters to the army guaranteeing to pay all liabilities and provide fire protection.

During the period of 1920 through 1929 Goddard wrote four reports that he also did not want to be made public because he felt the public would find his ideas difficult to accept. In these papers he set forth the principles of lunar and interplanetary flight and documented his interest and appreciation of the potential of rocket power. He pointed out the importance of photographing the moon and the planets, the use of gyros and flight-path correction by small rocket motors, the need for ablating heat shields on reentry, and the advantages of liquid hydrogen and oxygen as propellants. Goddard also considered man essential for landing upon and taking off from planets.

Among other ideas, he discussed the use of electric propulsion, a solar-powered generator using a mirror collector, and methods of producing an ionized jet of gas and accelerating it electrostatically.

For the first time he also mentioned manned space flight.

He suggested a 1,200-pound manned capsule as an observation compartment (the Mercury capsule, Freedom 7, used on our first manned space flight with astronaut Alan B. Shepard, weighed 2,000 pounds), and the "most economical acceleration" of about 4.8 g's. He also presented calculations on soft landing on the moon and the production of hydrogen and oxygen by solar energy on the moon and planets. "In the case of Venus," Goddard suggested, "it is very likely that the wind could be used as motive power, as there appears to be good evidence of strong winds."

At the time the newspapers were full of the flight of July 17, 1929, it happened that Harry F. Guggenheim and Charles A. Lindbergh were sitting by the fireplace at Falaise with Mrs. Guggenheim. The men were discussing what was the next area to be explored in aeronautics, when Mrs. Guggenheim interrupted, saying, "Look here, perhaps this is what you two want." The men were intrigued and Mr. Guggenheim asked Colonel Lindbergh to go up to New England and meet Dr. Goddard and look at his laboratory. This he did on November 23, and apparently the report was favorable, for a generous grant was made in June 1930 by Daniel Guggenheim, the father of Harry. Meanwhile, in December 1929, the Carnegie Institution in Washington advanced $5,000 to the Smithsonian for Goddard's continuing research.

Backed by a two-year $50,000 Guggenheim grant, Goddard embarked on a full-scale rocket-testing program. Taking a two-year leave of absence from Clark University, he and his assistants and equipment moved to the desert in Roswell, New Mexico, in 1930. They established their headquarters in a big Spanish house called Mescalero Ranch outside Roswell. They built a workshop and a twenty-foot static test stand nearby and another eighty-five-foot launch stand about ten miles northwest of Roswell on the prairie. The location was just right for year-round testing in a temperate climate—no fire marshals to complain and no people to frighten. Goddard's aim was to develop the many elements of a sounding rocket that he had earlier conceived but had not been able to build because of the lack of funds and facilities.

On December 30, 1930, he launched the first of his rockets from the New Mexico testing ground. It was eleven feet long, weighing 33.5 pounds without fuel. Fueled and in flight it reached an altitude of 2,000 feet and a speed of about 500 miles per hour, which was quite a contrast to the 184 feet reached by his first liquid rocket in 1926.

Although the first New Mexico firing reached a very high altitude, Goddard was concerned with the lack of precise flight control. The rocket had swerved and yawed as all other rockets had done before. He knew that there was a pressing need for controlling a rocket in flight. But by 1932, in the midst of the depression years, the Guggenheim funds were depleted and the second two-year grant had to be postponed. Goddard returned to Worcester and resumed teaching at Clark University. He again turned to Abbot at the Smithsonian, asking for $250 for tests aimed at reducing the weight of rocket parts. Abbot found the money for Goddard.

In 1934 the Guggenheim Foundation was again able to reopen the Goddard New Mexico "rocket range." Goddard immediately went to work on trying to perfect a rocket control system. When Goddard wrote to Abbot on September 4, 1934, that he had received funds from the Guggenheim Foundation, Abbot replied, "May I urge you to bend every effort to a direct high flight? That alone will convince those interested that this project is worth supporting."

On March 28, 1935, Goddard successfully fired a gyroscopically controlled rocket that achieved an altitude of 4,800 feet and a top speed of 550 miles per hour. Once again, on July 12, 1935, he successfully launched a rocket that climbed 6,600 feet in another completely stabilized vertical flight. Reporting this milestone in rocket development to Harry F. Guggenheim, Goddard stated, "Last Friday, we obtained

Harry F. Guggenheim, *left,* and Charles A. Lindbergh, *right,* during visit with Goddard in Roswell, New Mexico, in September 1935. (*The Smithsonian Institution*)

beautiful stablization. The rocket corrected itself, moving each side of the vertical, during the entire period of propulsion, fourteen seconds, reaching a height of about a mile and a quarter."

Satisfied with the results, Goddard then turned his attention to the problem of developing lightweight rocket parts and structure. When special problems arose, such as importing special equipment from abroad, it was always to Abbot and the Smithsonian that Goddard turned for help.

Fully appreciating the historical importance of his work, on November 2, 1935, Abbot asked Goddard to give a complete 1934 Series A design rocket to the Smithsonian. Goddard sent the rocket but asked that it not be displayed at that time. Goddard's wishes were respected. When it arrived, the box containing the rocket was bricked inside a false wall in the basement of the Smithsonian. After World War II, it was removed and placed on display.

Space travel reached a new high in scientific respect when The Smithsonian Institution published Goddard's second paper entitled "Liquid-Propellant Rocket Development" on March 16, 1936. Whereas his 1919 paper had concerned itself only with the theory of rocketry, the 1936 paper estab-

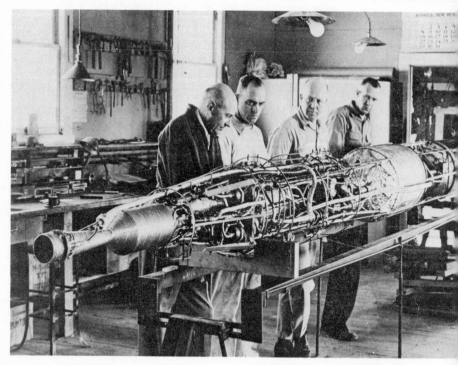

A liquid-fueled rocket with turbopumps, being worked on in Goddard's workshop in Roswell, New Mexico, on January 31, 1940. (*The Smithsonian Institution*)

lished his claim on the world's first liquid-propellant rocket flight. This paper also showed that Goddard was far ahead of all others in his field and presented indisputable facts about the existing future capabilities of a rocket.

Goddard remained in New Mexico testing rockets until World War II. Then again, as in World War I, Goddard gave his services to his country. He left his desert test center and in 1942 joined the Naval Engineering Experimental Station at Annapolis, Maryland, where he continued to work until his death in 1945.

Goddard's choice to do research in the field of rocketry caused him many frustrations and setbacks. He was content, however, to spend over half his life in rocket research, building and conducting untold numbers of rocket experiments, of which many were failures, trying to make each rocket go up just a little higher than the last one.

From the cold little Worcester firing range to the hot desert test site Goddard achieved tremendous advances in rocketry almost single-handedly. He made a mathematical analysis of multistaged rockets. He established that liquid fuels could be used in rockets. He developed the centrifugal propellant pump. He developed a rocket motor using liquid fuels and used it in a liquid-fueled rocket in 1926, nearly five years before the first German experimental flights and approximately twelve years before the Germans used it in the V-2.

He was first to shoot a rocket faster than the speed of sound. He developed the basic idea of the bazooka in 1918, at the end of World War I, and the weapon was actually used in World War II. He developed jet vanes and gyroscopic rocket controls ten years before the Germans did. He developed the mathematical theory of rocket flight and rocket propulsion on which all modern military and experimental calculations are based. He was the first to prove, both mathematically and then by actual test, that a rocket will work in a vacuum, that it does not need air or other substances to push against. In all, he was granted sixty-nine patents in his lifetime.

Like many other men of vision who have made valuable contributions to science, Goddard and his work were little known to the people of his own country during his lifetime.

Esther Goddard, in her recollections of her husband, says, "It is well to remember that this kind of life was lived while he taught a full schedule at Clark, and carried his full share of committee work. His precious research was for late afternoons, weekends, academic vacations, and above all the summertime. After fifteen years of working and waiting and hoping came the great adventure in New Mexico.

"During the fruitful years of full-time experimentation financed by the Guggenheim family, he was an extremely happy man, doing what he most wanted to do, with adequate funds in optimum surroundings . . . Sometimes he worked against exasperating odds—tests aborted, materials faulty, fuels impure, weather uncooperative. But he persisted in his dialogue with nature, asking questions, making trials, listening for verdicts, and building upon them. Then one day there would come a perfect flight: that was happiness indeed.

"My husband's dedication to one problem did not prevent his taking a part, as a citizen, in community life. Many invitations to speak to church and service groups were accepted, as part of his duties to the community. He voted in all local and other elections.

"Yet he was essentially a 'loner,' and the plight of those like him still worries the scientific and military world today. Each of the choices imposed by his shining goal brought one thing —unremitting work. Yes, Robert Goddard was a work horse, to the uninitiated observer. But I am sure the reader will understand that it was knowledge, hard won, that lifted away the drudgery, and left only glorious fulfillment . . .

"Such was the life my husband chose to live. Whether or not such a life could be created today, when the pace is much swifter, is being debated. The key to the full life, which is time for quiet thinking, is more difficult to attain, but surely it is not unattainable, even now.

"Some writers have tried to portray Robert Goddard as a martyr of sorts. This is far from being the case. His diaries record warm and friendly interest in his work from his peers who understood what he was trying to do. Among these, two men stand out—Dr. Charles G. Abbot, his closest friend and confident, and Harry F. Guggenheim, who also understood the problems of a pioneer. They showed a faith and encouragement that shines through their letters for several decades.

"Further, the record shows that Dr. Goddard received more financial support for his high-altitude nonmilitary rocket research than any other single scientist had been given for one project up to the time of World War II—and even in his own time he suspected that this was true.

"So shed no tears for Robert Goddard. He chose a field for his particular search for truth, and devoted his life unremittingly and joyfully to bringing his dream through hope into shining reality."

6

Hermann Oberth

It is reassuring to see that science and progress suffice to overcome national prejudice. I can think of no better way to thank the Société Astronomique de France than to pledge myself to work on behalf of science and progress and to judge people only on their personal merits.

Hermann Oberth, 1928

Hermann Oberth, the last living famous rocket pioneer, was eighty years old in June 1974 and lives at Feucht near Nuremberg, Bavaria. Although he has not worked on space travel since 1964, he is still the recognized father of German rocketry. On his seventieth birthday a group of his space scientist friends presented Oberth their congratulations in the form of a bound volume of well-wishing letters at the annual meeting of the Hermann Oberth Rocket Society in Huntsville, Alabama.

Ernest Stuhlinger, in the preface of this volume entitled

Hermann Oberth. (*The Smithsonian Institution*)

From Dreams to Reality, wrote, "Your first birthday present, no doubt, is the fact that man has accomplished the flight through satellite orbits, that manned lunar flight is imminent, and that travel to Mars and Venus will come within reach, hopefully, during your lifetime. You were witness and actor in all the major phases of this fabulous evolution."

Hermann Oberth, a schoolteacher turned space scientist, was born on June 25, 1894. His family was of German descent but lived in the small city of Sibiu, on the northern slopes of the Transylvanian Alps, at the time a part of the Austro-Hungarian Empire. Sibiu and other nearby towns, however, had been settled in about the twelfth century by a group of Saxon Germans. They continued to call their town by its German name, Hermanstadt, and maintained their own culture and customs throughout the centuries. Oberth's father was a medical doctor at the Franz Joseph Hospital. Two years after Hermann was born they moved to Sighisoaro.

Oberth attended the local elementary school and was an avid reader. When he was about twelve years old, he was deeply impressed with Jules Verne's *From the Earth to the Moon,* the story of how three men were fired out of a huge cannon in a spacecraft shaped like a cannonball that flew around the moon, missing the moon and eventually landing back on earth.

From that point on, Oberth began to be preoccupied with space flight and the idea of a journey to the moon. He mulled over details of Verne's stories, such as the plan to use retro-rockets for slowing the cannonball's speed before landing and the operation of the rocket on the moon where there is no air, and became convinced that Verne was correct. He was sure a rocket could work in airless space.

The one part of Verne's story that Oberth just could not bring himself to accept was that the passengers in the cannonball spaceship were protected from shock during launch by the use of a water-filled cushion. Oberth, who was very capable in physics, decided to make some calculations of his own. He estimated that the passengers would actually need a cushion of about 1,000 miles thick to avoid being flattened by the launch acceleration.

These calculations convinced Oberth that for space flights to become practical a rocket system had to be developed that would have very low acceleration at launch and then gradually accelerate to its maximum speed of seven miles per second. He calculated the spacecraft would also need a braking apparatus, or retro-system, to slow it to a safe speed for landing on the moon.

With his consuming interest in space flight and the adventures of *From the Earth to the Moon* to inspire him, Oberth found it difficult to think about the medical school that his father wanted him to attend. But his objections were short-lived, and he did enter the University of Munich as a medical student.

In addition to his required medical courses the school permitted Oberth to enroll in two elective courses—mathematics and astronomy. Some say he also studied rocketry with a private tutor. Thus, it seems that while in medical school rocket flight was still one of his major interests and that he was also preparing himself in scientific fields that would be directly useful in advancing his space studies.

Oberth was twenty and still a medical student when World War I began. Austria-Hungary and Germany were allies, and almost immediately Oberth found himself inducted into the army as an infantryman.

His military career in the infantry ended abruptly when he was wounded. After recovering he was transferred to the No. 22 Field Ambulance Unit. Since his duties were not in a very active combat zone, he saw fairly few casualties. Yet after the war Oberth commented, "This was a piece of good luck, for it was here I found that I should probably not have made a good doctor."

The inactivity of his hospital assignment gave Oberth time to think about space and space flight problems. In fact, he spent the remaining war years trying to work out a mathematical theory for space flight. It did not take him long to realize that he could not work on his space flight ideas until he went back to school and mastered more mathematics, astronomy, and physics.

Just before the end of the war Oberth married Mathilde Hummel, a Transylvanian of German descent like himself, who has remained his companion through all his years in school and throughout his career.

Oberth resumed his studies when the war was over, first in Munich and then later at the University of Heidelberg. He completely dropped all pretense of a career in medicine, and his courses were now selected to help him with his teaching career and his interest in space flight. In addition to his teaching courses he took mathematics, theology, chemistry, meteorology, astronomy, and applied physics.

The year 1922 was the turning point in Oberth's career. Although he was studying to be a teacher in mathematics and physics, he submitted a doctoral thesis on an unusual topic. The title of the paper was "By Rocket to Interplanetary Space." He tried to interest scientists of the day, but both engineers and scientific experts turned him down, one after the other. It was then, in desperation, that he decided to force their attention by arousing popular interest. For years Oberth had no choice but to use this means to draw attention to the possiblity of space travel.

He gave his thesis to a friend, Dr. Paul Requadt of Hannover, who submitted it to over twenty publishers. It is doubtful that many of them read more than just the title, but if they did, they quickly rejected it because of the many pages of mathematics. Requadt decided to try once more and sent the manuscript to Rudolf Oldenbourg, a Munich publisher. At first Oldenbourg was going to reject it, since he was not interested in science fiction, but on a more thorough examination of the contents he was impressed with the confidence with which the unknown author had stated his belief that space travel was possible and that it would be developed in the very near future. Oldenbourg wrote to Oberth that he would publish the book if Oberth would pay the publication costs.

By Rocket to Interplanetary Space was published in 1923. The book was the real beginning of Oberth's space career and set the pattern for almost forty years of his working life.

The published work was somewhat different from Oberth's original scientific thesis on rocketry. In his desire to interest other engineers and scientists in his theories he included material that would appeal to the general reader, thinking that if the public became interested, the scientists who had ignored his work might reconsider and examine his work professionally. The title certainly had popular appeal, and the first printing was sold out. In fact, advance orders almost exhausted the second edition when it appeared two years later.

The manuscript was under 100 pages, yet within those pages Oberth had developed his space flight theories in great detail. He explored the possibility of sending a vehicle into space, concluding that the craft would have to be launched by a rocket, since that was the only type of propulsion that could operate in outer space. He also asserted that a spacecraft could be made large enough to carry human passengers and safe enough for the passengers to survive the journey.

The book was written in three sections. The first section was a technical discussion of how a rocket could really travel faster than its own exhaust and would continue to rise higher even after its propellants were exhausted and its engines shut off. The second section described with diagrams the details of a high-altitude research rocket. But it was the third section that interested the general reader most, because of the nontechnical writing style and the subject matter.

In this section Oberth pictured a space station as traveling in orbit around the earth, much like our Skylab, studying the weather conditions on the earth and serving as a space refueling depot for rockets traveling even deeper into space.

He described the spacesuit, the special seat in the spacecraft, and the special shoes that would enable a space traveler to walk in the spacecraft under zero gravity conditions. In this part of the book Oberth made it sound as if science was momentarily ready to embark on space voyages.

Some of the things he wrote about were previously known facts, but many were original. In fact, three other rocket pioneers—Goddard, Tsiolkovsky, and Esnault-Pelterie—were independently working and thinking about similar space flight problems.

Although these men did some of their most fruitful work in rocketry at approximately the same period in time, they had very little contact with one another. At one point, however, in spring 1922, Oberth and Goddard did correspond. At that time Oberth was in the final stages of editing his book and was startled to read in a newspaper that an American, Robert H. Goddard, had published a booklet on rocketry through The Smithsonian Institution entitled "A Method of Reaching Extreme Altitudes."

Oberth tried unsuccessfully to obtain Goddard's paper in Heidelberg, and on May 3, 1922, he wrote to Goddard in Worcester, Massachusetts. The letter was very brief and to the point, stating that he, Oberth, had been working on rocket problems for many years and had been in the process of publishing his results when he read about the Goddard publication. He went on to say that he had not been able to find the booklet in Germany and asked Goddard to send him a copy. He promised to send Goddard a copy of his publication as soon as it became available and concluded with "I think that only by common work of the scholars of all nations can be solved this great problem."

Goddard responded promptly, but unfortunately Goddard's copy of the letter to Oberth was lost, and the original was destroyed with other of Oberth's papers during World War II. Oberth eagerly read Goddard's booklet from cover to cover studying in particular Goddard's calculations on how much solid propellant would be necessary to lift rockets of various sizes from the earth to the upper atmosphere. Oberth agreed with everything in the book except Goddard's calculation of how much flash powder was needed to cause an explosion on the face of the moon that would be visible on earth.

After reading Goddard's statements Oberth may have felt that in some respects he was more advanced than the American in his concepts. After all, Goddard's space flight proposal at that time only involved making a flash on the moon's surface whereas Oberth was submitting plans for a manned spacecraft journey to the moon. In reality this was not the case, and if adequate communication between the two men could have been established, Oberth would have learned that Goddard's thoughts had gone much beyond the flash-powder proposal.

Oberth received a great deal of fan mail about his book, but the experts remained uncomplimentary. Most scientists, fearful of being labeled cranks, ignored him. Their attitude hurt Oberth's career as a scientist for many years. His critics pointed to the book as proof that a man with such a wild imagination should not be taken seriously.

One technical journal reported, "we believe that the time has not yet come for delving into such problems as these and indeed probably never will come." In another journal a reviewer disputed Oberth's claim that a rocket could function in a vacuum. And the worst was yet to come. In 1925 Carl Barthel, a local banker who had offered to finance Oberth's rocket experiments, dropped his support. The banker's decision was based on a report by Professor Rudolf Franke of the Charlottenburg Technical High School prepared at Barthel's request. Franke stated that Oberth's assumptions were wrong and that his basic calculations were incorrect. Oberth's plans and drawings were returned, and he was never told where the two scientists' opinions differed. At that time Franke's expertise was above question.

Oberth replied to his critics' attacks in the preface to his second edition, writing, "In the third part of my treatise I make some fantastic assertions which, although of a kind not usually found in scientific works, cannot be refuted on scientific grounds; it must be remembered that in this third section we are dealing with a special situation."

Unable to win over his critics, Oberth agreed to an offer from Max Valier, a builder of rocket cars and author of scientific articles, to rewrite his book in an even more popular vein. Oberth hoped to reach still more people and eventually arouse the interest of the scientific world.

Working together, the two men amused themselves in figuring out the dimensions of a moon-cannon such as Jules Verne had described. Their calculations showed that the gun would have to be about 3,000 feet long and be built inside a mountain near the equator in order to get the tip of the muzzle about three miles above sea level.

Although his experience with Valier temporarily took the sting out of the criticism directed at him, he left Germany before the new version of his book was released. He returned to the town of Mediash in Transylvania and for the next few years taught mathematics and physics.

Valier's version of Oberth's book was published in 1924, entitled *A Dash into Space*. The book did not translate Oberth's ideas faithfully, and in an attempt to be entertaining Valier used only parts of Oberth's scientific material. What he did include was presented inaccurately. Some of the material was printed in smaller type, encouraging the reader to skip over Oberth's complex ideas and formulas. Despite these changes the book had a good steady sale and certainly contributed to the growing interest in rocketry and space travel.

In addition to Valier's attempts another contribution to the growing interest in space travel was the publication of a small book entitled *The Possibilities of Reaching Outer Planets* by an engineer named Dr. Walter Hohmann. His eighty-three-page book was restricted to mathematical aspects of space travel but nevertheless was read by a growing number of young space enthusiasts.

It was most fortunate that one of Oberth's fans and a reader of Valier's book was Willy Ley, who later became an internationally known writer on space matters. In 1926 Ley published a book called *Travel in Space,* the first book that really could be read and understood by a layman without too much difficulty. Ley had succeeded in writing a book that made Oberth's pioneering theories accessible to a wide audience.

On June 5, 1927, an important event in German rocketry took place. A handful of men who had read Oberth's, Valier's, and Hohmann's books and believed enthusiastically in the future of space flight formed the Society for Space Travel (Verein fur Raumschiffahrt), better known as the VfR. The group immediately asked such well-known rocket men as Hermann Oberth and Robert Esnault-Pelterie to join, and they accepted.

The time for interest in space travel had come. Almost simultaneously in other countries scientists were forming groups to exchange ideas on astronautics. Only three years before the formation of the VfR, Russian scientists with similar interests discovered the pioneering work of Tsiolkovsky and others and convinced their government to support the special Central Bureau for the Study of the Problems of Rockets (TsBIRP) and the All-Union Society to Study Interplanetary Communications (OIMS).

In the United States G. Edward Pendray and his wife, the late Leatrice Gregory Pendray, founded the American Interplanetary Society, later to become the American Rocket Society and in 1963 to merge with the Institute of Aerospace Sciences to form the American Institute of Aeronautics and Astronautics.

In France, three days after the VfR was organized, Esnault-

Pelterie gave his first space lecture before the French Astronomical Society.

With the absolute support of the VfR, whose members he had clearly inspired through the years, Hermann Oberth had finally become accepted in scientific circles as a rocket expert. He was even able in 1928 to appear before the distinguished meeting of the Scientific Society for Aeronautics to defend his work against the severe criticism of Professor Hermann Lorenz, whose comments had been published in the journal of the Society of German Engineers. After Oberth's effective and concise speech, Lorenz admitted he had not actually read Oberth's work and never again spoke publicly about space travel.

In 1928 Oberth was elected president of the VfR, which had grown from a dozen or so charter members to over 500 in just one year. His articles were in demand now, and he contributed an essay to a collection entitled *The Possibility of Space Travel* edited by Willy Ley. Oberth's first publisher, Rudolf Oldenbourg, printed the anthology. Oldenbourg also asked Oberth to begin work on an expanded version of his famous book *By Rocket to Interplanetary Space*. At that point Oberth's work had become so well known that he received an invitation from the movie director Fritz Lang to come to Berlin and serve as the technical adviser on what was to be the first movie about space travel, *A Girl in the Moon*.

Fritz Lang, one of the most famous directors of his day, first thought of making a motion picture about space travel while he was directing a futuristic film called *Metropolis*. The idea became more attractive as he collected all of the available literature on space travel. In particular he was impressed with Oberth's description of the craft designed to carry a crew of space travelers to the moon. The picture, a romantic fantasy, was actually based on Oberth's book. Lang, wanting as much realism as possible, asked Oberth to direct the building of the moon ship. Recognizing the educational value of the picture and its possible benefits to rocketry itself, Oberth took a leave of absence from his teaching job and went to Berlin.

Life in Berlin was a new experience for the teacher from Transylvania. Overnight Oberth was in the center of the strange and glamorous world of movie production—the people, their specialized language, and their rapid pace of life. It was all very different from the quiet country life he was used to. In addition to adjusting to a new environment the job of designing the movie set was very demanding.

In the midst of all this, Oberth was faced with an unexpected monumental task. Willy Ley arrived in Berlin with a proposal from the VfR that the film company finance Oberth's development of a liquid propellant rocket. The VfR pointed out that for about the same amount of money that Lang was spending on the picture's publicity, a workable liquid propellant rocket could be built, which in turn would provide good publicity for the motion picture studio. Oberth, not fully aware of what he was getting himself into, approached Lang with the plan.

Lang thought the idea was brilliant and asked Oberth how much he needed to do the job. Oberth had no idea of how much it would cost to build a rocket, for he was a man of theory and had never constructed anything. To get the funds Lang had to convince the financial management of UFA Film Company for which he was making the picture. The management, however, was not convinced that building a real rocket would in any way benefit the picture they were producing. But the enterprising Lang was not discouraged and finally offered to donate a considerable amount of money if UFA would match his offer. The company agreed to this compromise.

Oberth's schedule was now impossibly heavy. He had to design the rocket and the spacecraft to be used in the movie, as well as supervise the building of realistic scenery and the

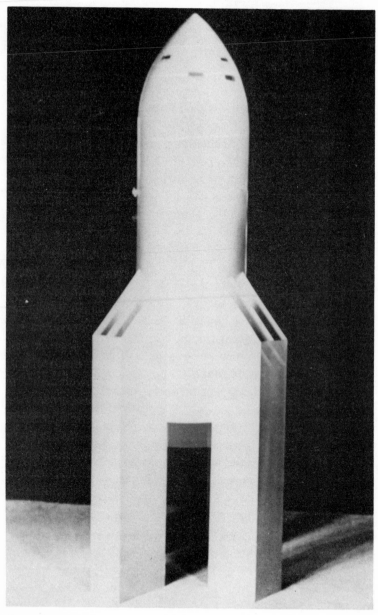

A model of the rocket ship Oberth designed for the film *The Girl in the Moon*. (*The Smithsonian Institution*)

construction of a spaceship cabin with all of the very complex dials, lights, and levers.

In addition to all this activity he had the responsibility of building a workable liquid propellant rocket in time for the premiere of the movie. Although he knew very little about actually building such a rocket, Oberth was expected to produce it in three months. The task was impossible under the best conditions. Oberth was a brilliant mathematician, but he did not have the staff or the expertise needed to build a rocket in even double the time allowed. Nevertheless, Oberth, caught up in the enthusiasm of the day and anxious to oblige the VfR, advertised for help in the newspapers.

His first applicant was a man named Rudolf Nebel, a wartime pilot with eleven enemy planes to his credit. Nebel did have an engineering degree, but because of the war he had never held an engineering job. Oberth's second applicant was a Russian by the name of Aleksander B. Shershevsky who had been sent to Berlin by the Soviet government to study aircraft engineering. Shershevsky found life in Berlin so interesting that he had overstayed his time and was afraid to go home. Oberth never did like Shershevsky and described him as "the second laziest man I had ever met; he had very bad teeth and never enough money to have them put right. If I lent him some to go to the dentist, he forgot his teeth and spent all the money on pleasure."

As Oberth quickly learned, neither man had the engineering experience that he himself also lacked. As a result, mistakes were made from the very beginning. The three men, having set an unrealistic three months as their goal of completion, designed a rocket that was a much smaller version of the V-2 rocket (which a decade later was to be used against England during World War II). Oberth calculated that this rocket would be able to reach an altitude of about twenty-five miles.

The propellants they selected were gasoline and liquid

oxygen, since they were inexpensive and easy to obtain. At that time no information existed in Germany on the combustion of this propellant combination, so Oberth had to start from scratch to experiment with them. He suspended several pints of liquid oxygen in a container in the center of his workshop and directed a thin stream of burning gasoline into it. After a few tense moments he saw with relief that the propellant combination burned well and did not explode as others had predicted.

On the basis of this experience he next poured a larger quantity of gasoline in a layer over the surface of the liquid oxygen and lit it. This produced a violent explosion that threw him across the workshop. When he recovered consciousness, he could not see and was barely able to hear. The damage severely harmed his eyes and perforated one eardrum. Although his eardrum partially healed, he did lose the sight in his right eye.

Oberth repeated the experiment as soon as he could, but with extreme care, and was able to learn how the propellant combination really burned. This knowledge allowed him to reduce the size and weight of the rocket engine, as well as the whole rocket. In Germany Oberth's design of the combustion chamber and cone nozzle was a major technical advance in rocketry.

Eventually Oberth dismissed Shershevsky and, working with Nebel, prepared for the first firing of the engine in the rocket's framework. The people who were there to witness the test expected to see it explode, but to everyone's surprise the engine fired exactly as he had predicted.

Despite some small successes time was running out. The premiere was drawing nearer, and Oberth still had a long way to go. He and Nebel worked around the clock, often up to sixteen hours a day. But this was not enough. Oberth had undertaken too much, and, shattered physically, one morning he did not appear at the laboratory. He had the good sense

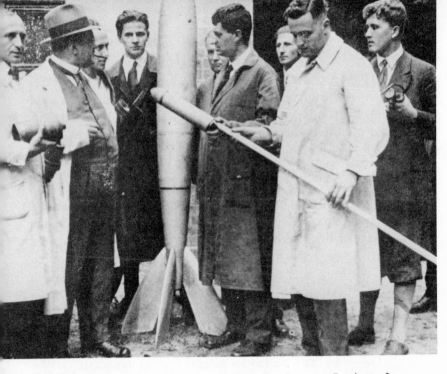

A photograph taken in August 1930 at the German Institute for Chemistry and Technology; Hermann Oberth, *standing to right of upright rocket,* flanked by a group of his colleagues, including Wernher von Braun, *far right,* as a young man. (*NASA photograph*)

to quit and return to familiar surroundings to recover. He went back to the security of his family and the mountains.

The Girl in the Moon had a spectacular premiere on October 5, 1929, without its much publicized rocket launching. But it made little or no difference to the movie-going public. The picture was a huge success, and the public soon forgot the rocket incident.

The project, however, was not a complete failure, since some progress had been made by Oberth on various tests. After a complete recovery he returned to Berlin and with the help of friends and the VfR finished the rocket. It was not launched, but Oberth did manage to get the German Institute for Chemistry and Technology to test his little rocket motor. It performed successfully, and he probably could have raised funds for continued research, but he was called back from his leave of absence to his teaching post at Mediash. Since Oberth had a family to support, he left Berlin, but his colleagues at the VfR continued building and testing liquid propellant rockets.

It was not until eight years later, in 1938, that he resumed his experiments as part of a rocket research program at the College of Engineering in Vienna. He was forty-four years old then and eager to work again in the field he had pioneered. But not long after arriving in Vienna he found himself in a very difficult situation. Austria had become an ally of Nazi Germany, and the German army and the Luftwaffe (air force) were competing to become the favored branch of Hitler's war machine. Both sides had become interested in rockets.

The army had started its research at Peenemunde, a new rocket installation, and the Luftwaffe organized its own group of rocket experts in Vienna, which included Oberth. Although the scientists had virtually no funds or equipment to work with, the Luftwaffe kept them on its payroll so that no one else could use them. Oberth was given one mechanic and a small test pit to carry on his work. The situation became intolerable, particularly when he was informed that his superiors were disappointed with his progress. Then quite unexpectedly he was transferred to the College of Engineering in Dresden and was instructed to develop a fuel pump.

Oberth had no way of knowing that this assignment too was a fraud and that he had been transferred to the army without his knowledge. The army wanted him in Dresden for a period so that they could check his loyalty before assigning him to the secret laboratories in Peenemunde.

When Oberth protested that it was impossible to develop the pump under the existing conditions, he learned that a pump had already been built. Indignant and disappointed,

Oberth requested permission to go home. The Gestapo told him this was impossible because he had accumulated too much secret information about German armaments. Furthermore, he was made to understand that he had to become a citizen of Germany or go to a concentration camp. Under pressure he became a German citizen. When the papers were finally prepared for his transfer, the scientists at Peenemunde, under the leadership of Gen. Walter Dornberger and the rocket enthusiast Wernher von Braun, had just about completed the V-2 rocket. He arrived much too late to take part in building the weapon. Again at Peenemunde Oberth felt useless and ignored.

In 1943 Oberth was transferred to Wittenberg, where he was supposed to build a solid-propellant rocket using ammonium nitrate for antiaircraft purposes. But in spring 1944 the chemical works in Germany were heavily bombed, making it impossible to obtain the necessary ammonium nitrate.

When the Allies entered Germany at the end of the war, Oberth was arrested and put in an internment camp for a short period. The chemical plant where Oberth had been doing his research was located in western Germany, which fell into British hands. By the time he was released, a number of other German rocket men had been sent to the United States. He was not included in the group and returned home in 1947.

Conditions at home were very difficult after the war, and Oberth was not able to find any form of work. His family situation had also changed. His daughter Erna had graduated with a law degree in Bern and had become a legal adviser in Nuremberg; his son Adolf was studying for his Ph.D. in chemistry at Munich; his eldest son, Julius, who would have been twenty-six, was missing in Russia; and his daughter Ilse had died in active service in 1944.

Finding nothing to keep him at home, he left, crossing the border illegally, and entered Switzerland. For almost a year he lived as a consulting engineer and for another year as a guest of a fireworks manufacturer by the name of Hans Hamberger. Then the Italian navy engaged him to complete the solid-propelled antiaircraft rocket that he had begun in Wittenberg. He was paid 120,000 Swiss francs and was sent to La Spezia, where his wife joined him. He continued his work assisted by three Germans and five Italians in comfortable and pleasant surroundings. He could not convert his ideas into a workable rocket, however, and the project failed. In spring 1953 Oberth returned to Germany to his old teaching position and started work on a new book called *Man into Space*.

In 1955 Wernher von Braun, while working for the United States army developing short-range missiles, asked Oberth to join him. Oberth accepted and again left his homeland, this time for the Redstone Arsenal in Alabama. Oberth worked as a consultant to the army team in Alabama from 1955 to 1958 and again in 1961.

The white-haired Oberth and his wife now live on a modest monthly pension in an attractive two-story house in the Bavarian village of Feucht, near Nuremberg.

Mustached and bespectacled, Oberth still has the vivid imagination of his Jules Verne boyhood. Oberth believes that unidentified flying objects have been sighted for thousands of years, "the first being sighted in North Egypt, over 3,500 years ago." He also feels that man will someday travel at the speed of light—a speed that many scientists claim cannot be attained. What greater goal can the man who led Germany's first efforts to conquer space leave to the future generations of rocket pioneers?

7

Pioneers of the Future

In no case must we allow ourselves to
be deterred from the achievement of
space travel, test by test and step by
step, until one day we succeed, cost
what it may.
Robert Hutchings Goddard, 1933

The legacy handed down by the famous pioneers of rocketry
—Congreve, Tsiolkovsky, Esnault-Pelterie, Goddard, and
Oberth—and two later generations of rocket and spacecraft
builders is rich in dedication, accomplishments, and promise
for the future.

Their combined efforts have made it possible since 1957
for man to send unmanned vehicles into space, into orbit
around the earth and the moon, to the moon, and to the
nearby planets of Venus and Mars. Then on March 2, 1972,
man achieved another major breakthrough with Pioneer 10,
which launched from Cape Canaveral and reached the orbit
of Mars on May 25, 1972. Once beyond the orbit of Mars
Pioneer 10 sailed into the unexplored region of the asteroid
belt, undamaged by the innumerable fast-moving bits and
pieces of matter hurtling through space.

On that spectacular journey Pioneer 10 flew by the planet
Jupiter in December 1973, photographing it and recording a

Wernher von Braun demonstrating a wheelchair operated by a "sight switch" that translates eye movements into movements of the wheelchair by a complex system of relays, switches, and motors—just one example of space-age technology being used to aid the handicapped. (*NASA photograph*)

great amount of data. During its brief journey past the planet it encountered and survived radiation fields strong enough to kill every man, woman, and child on earth. Scientists say that the only reason Pioneer 10 was not destroyed was that it flew though the radiation at a speed of about 96,000 miles per hour.

After it has traveled for about fifteen years, by 1987 this amazing unmanned spacecraft is expected to cross the orbit of Pluto, totally escaping our solar system and then journeying forever among the star systems of the Milky Way galaxy—a feat that even our visionary pioneers would have thought impossible so shortly after the birth of the space age.

But the first two decades of space flight have not been limited to unmanned space exploration. Man has also flown into space, remained in orbit around the earth for increasingly longer durations, and even landed on and returned from the moon.

At the same time that scientists have been sending their complex machines and highly trained astronauts and cosmonauts into space in search of new knowledge, they also have applied the knowledge gained from the billions of dollars spent on research to improving the daily life of mankind generally. These improvements can be found in communication (such as worldwide television), weather forecasting, navigation for ships and aircraft, protecting ecology on a worldwide basis, and searching for new resources and protecting existing ones. Perhaps the most dramatic spin-off benefits from space technology have occurred in the development of electronic devices. The list of direct and indirect benefits is almost endless.

A great deal of progress has been made, but it is important to remember that these accomplishments are not an end in themselves but just the beginning. Therefore, the challenge to the young people reading this book—the future rocket pioneers—is even greater than it was for our pioneering ancestors.

These newer challenges will have their share of rewards and frustrations. One lesson that we can learn from history is that in any field of endeavor we are apt to find at some time or another the road to progress being blocked and ofttimes stopped by the uninformed and informed alike. People often do not understand or appreciate the value of a new idea and its eventual impact on their way of life. The tendency has always been to see only the immediate problems, to the detriment of the future.

A typical example are the problems of the environment. Shortages of food and natural resources, among others, were discussed several decades ago by the well-known conservationist Fairfield Osborn in his book *Our Plundered Planet*. Looking to the not too distant future, he wrote, "The tide of the earth's population is rising, the reservoir of the earth's living resources is falling. Technologists may outdo themselves in the creation of artificial substitutes for natural subsistence, and new areas, such as those tropical or subtropical regions, may be adapted to human use, but even such resources or developments cannot be expected to offset the present terrific attack upon the natural life-giving elements of the earth."

This vision and warning went unnoticed or was disregarded for a number of years. There are many other such examples of major contributions being initially rejected for lack of vision. The jet engine was turned down by American, German, and British engine manufacturers, and in each country it had to be developed by people who had no real experience with it; Kodachrome was developed by two musicians, following a procedure that Eastman's experts said would not work; the Accutron watch design was rejected by the entire Swiss industry; the polaroid photographic process was rejected by the established photographic industry; Xerox was rejected by the people in the reproduction-process business; and the "float glass" method for producing polished glass was rejected by the entire glass industry.

Rocket pioneers of the future can pursue careers in one of the established rocket fields or in new fields that are just emerging, or they can elect to pursue independent investigations into areas that today are undefined.

The tasks ahead are difficult and many. Much can be done to improve our earth technology satellites (communication, weather, earth resources, and so forth); Skylab has proven that man can remain in orbit for long durations, opening the way for building much larger and much more efficient stations in the future; the moon is now accessible to exploration; and Pioneer 10 has paved the way for man to follow into the solar system.

Imagine what might have happened if the four contemporary rocket pioneers—Goddard, Esnault-Pelterie, Tsiolkovsky, and Oberth—had been able to work together in one laboratory at the same time, exchanging ideas, solving problems, questioning approaches, and most importantly not wasting their time duplicating each other's efforts but driving rapidly ahead from one idea and concept to another. What great discoveries would they have made jointly? Would their team achievements have been accepted more readily than were their individual efforts?

How much sooner would man have conquered space? It is difficult to say, but probably much sooner. It seems probable too that our approach to rocket flight would not have been based on the use of expensive and wasteful disposable boosters. These four pioneers, with their combined range of experience, might have agreed that the best approach would have been to develop the less expensive space shuttle first, completely bypassing the throwaway booster.   .

We can only guess now about what might have resulted from such a team effort. Of one thing we can be sure, however; they would have done it better, sooner, less expensively, and with a greater degree of acceptance by the world at large.

Our first steps leading to cooperation in space came about

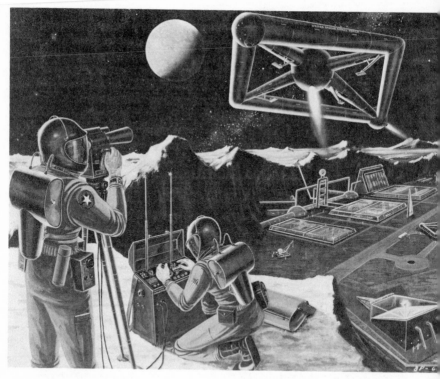

A preview of the future, when the moon may be used for the development of new energy sources. (*Martin-Marietta Corporation*)

as a result of self-preservation. Both the United States and the Soviet Union wanted to develop their respective capabilities to rescue their men from orbit. Perhaps the next step might lead to manned multinational flights in near-earth orbit and still much later to the moon.

It is not too difficult to picture that these steps, if successful, could lead to the establishment of an earth space station that would be used by all nations of the world and could eventually be used for the exploration of our solar system on a cooperative basis.

There is reason to believe that people of the future will

live on other celestial bodies. Looking beyond the conquest of our solar system, it is not unreasonable to picture humans living on other planets around other stars, building new societies on distant and unknown planets. We have broken our earthly bonds and are about to leave our natural earth spaceship and embark on voyages in man-made spaceships to the planets—perhaps even to the stars!

The future is in the hands of tomorrow's pioneers—those who will invent and build space-conquering machines and those who will fly them to explore and inhabit the universe.

Milestones in Rocketry

673   Discovery of "Greek fire," a mixture of pitch and naptha.

1232   Battle of Kai-fung-fu. Mongols led by Ogdai, son of Genghis Khan, repulsed by first reported use of Arrow of Flying Fire and Heaven-Shaking Thunder.

1687   Sir Isaac Newton develops his three laws of motion, the third law stating, "For every action there is an equal and opposite reaction," which established the principle of the rocket's operation.

1761   Haidar Ali, Prince of Mysore, defeats British forces at the Battle of Panipat with a rocket corps of 1200 men.

May 20, 1772   William Congreve, England's rocket pioneer, born in London, England.

1780   Tippoo Sahib, Haidar Ali's son, with a rocket corps of 5,000 men, defeats the British forces in battle several times near Seringraptam, India.

1805   William Congreve accompanies Sir Sidney Smith on the first naval rocket attack against the French city of Boulogne. About two hundred rockets are launched but attack fails due to high seas and winds.

1806	Second rocket attack is launched against port city of Boulogne. Attack successful. City burned and invasion fleet destroyed.
1807	British bombard city of Copenhagen with Congreve's rockets. City razed and Danish fleet destroyed, denying its use to Napoleon.
1813	British use Congreve's rockets against Napoleon's armies at Danzig and Leipzig.
Aug. 24, 1814	British use Congreve's rockets against American troops at Bladensburg, Md. American troops defeated, permitting capture of Washington, D.C.
Sept. 17, 1857	Konstantin E. Tsiolkovsky born in small town near Moscow.
Nov. 8, 1881	Robert Esnault-Pelterie born in Paris, France.
Oct. 5, 1882	Dr. Robert Hutchings Goddard born in Worcester, Mass.
June 25, 1894	Hermann Oberth born in Sibiu on northern slopes of Transylvanian Alps.
1903	Tsiolkovsky publishes his first paper on space travel, "Investigating Space with Reaction Devices."
1914	Goddard is granted a United States patent for a rocket using liquid and solid fuels and another for a multistage rocket for reaching high altitudes.
1918	Goddard successfully demonstrates his bazooka-type rocket at the Aberdeen Proving Grounds.
1919	Smithsonian Institution publishes Goddard's first paper, "A Method of Reaching Extreme Altitudes."
Mar. 16, 1926	Goddard successfully launches the first rocket powered by liquid propellants. The rocket traveled a distance of 184 feet in 2.5 seconds.
1927	Formation of the German Society for Space Travel (Verein für Raumschiffahrt).
1927	Esnault-Pelterie lecture before French Astronomical Society provides basis for his classical book *L'Astronautique*.
1928	Esnault Pelterie and André Hirsch cosponsor the REP–Hirsch International Astronautics Prize. First recipient Hermann Oberth.

1930	Formation of American Interplanetary Society by G. Edward Pendray and David Lasser. In 1934, the name was changed to the American Rocket Society.
Dec. 30, 1930	Goddard launches the first of his rockets from New Mexico testing station. Rocket reached an altitude of 2,000 feet and a speed of about 500 mph.
Mar. 28, 1935	Goddard successfully fires a gyroscopically controlled rocket to an altitude of 4,800 feet at a speed of 550 mph.
Mar. 16, 1936	Smithsonian Institution publishes Goddard's second paper, entitled "Liquid-Propellant Rocket Development."
July 1, 1939	Rocket Research Project is formed under Dr. Theodore von Karman at California Institute of Technology. This project became the nucleus of the nation's first center devoted to the research and development of rocket propulsion systems.
Oct. 1939	German scientists successfully fire and recover A-5 development rockets with gyroscopic controls and parachutes, attaining altitude of 7½ miles and a range of 11 miles.
July 1941	U.S. Navy initiates development of Mousetrap, ship-based 7.2 inch mortar-fired bomb, which became first USN rocket placed into fleet action, May 1942.
June 13, 1942	First test of the German A-4 (V-2) rocket at Peenemunde is unsuccessful.
Oct. 3, 1942	First successful launch and flight of the 5½-ton German V-2 at Peenemunde. Rocket travels 120 miles.
May– June 1943	Germans operationally test-fire 750 V-2s from Blizna, Poland, launching as many as ten a day.
Aug. 17–18, 1943	Royal Air Force attacks Germany's Peenemunde Rocket Research Center causing heavy damage and delaying V-weapon program by months.
June 13, 1944	The first German V-1s are launched from France against England, with four of the eleven striking London.
June 22, 1944	U.S. Army Ordnance awards a contract to California Institute of Technology for research and engineering on long-range rockets and their launching equipment.

Sept. 8,   First German V-2s fall on England.
1944

Jan. 24,   Germans successfully launch A-9, a winged prototype
1945   of the first ICBM (the A-10), designed to reach North
            America. A-9 reached a peak altitude of nearly 50
            miles and a maximum speed of 2,700 mph.

Mar.   Project Paperclip to recruit German scientists is initiated
1945   in the Pentagon.

Sept. 20,   Wernher von Braun and six other key German rocket
1945   scientists arrive in the United States under Project
            Paperclip.

Jan. 16,   U.S. upper-atmosphere research program is initiated
1946   with captured German V-2 rockets.

Mar. 22,   First American rocket to escape earth's atmosphere, the
1946   WAC reaches 50-mile height after launch from White
            Sands Proving Ground.

June 28,   First V-2 rocket fully instrumented by NRL for upper-
1946   air research is launched from White Sands Proving
            Ground and attains a height of 67 miles.

Jan. 23,   Telemetry operates successfully for the first time in a
1947   V-2 firing.

May 22,   The first Corporal E round is fired at White Sands Prov-
1947   ing Ground. This was the first launching of a surface-
            to-surface missile guided by radar ground control.

June 20,   Army Ordnance establishes the Bumper Project for de-
1947   velopment of a two-stage missile (German V-2 and
            modified WAC Corporal).

May 13,   A Bumper-Wac fired at White Sands Proving Ground
1948   is the first two-stage rocket to be launched in the West-
            ern Hemisphere.

July 26,   Two separate rockets are fired from White Sands, one,
1948   a V-2 reached an altitude of 60.3 miles; the other, a
            Navy Aerobee, reached an altitude of 70 miles, carried
            cameras that photograph the curvature of the earth.

Feb. 24,   Bumper-Wac No. 5 sends its upper stage to a height
1949   of approximately 244 miles and a speed of 5,510 miles
            per hour, the greatest velocity and altitude yet reached
            by a man-made object.

May 19, 1950   The first Hermes A-1 test rocket is fired at White Sands Proving Ground.

Dec. 1950   Construction starts at Grand Bahama Island for the first tracking station on the Florida Missile Test Range, later the Atlantic Missile Range.

Jan. 16, 1951   Air Force establishes Project MX-1593 (Project Atlas), study phase for an intercontinental missile. Contract was given Consolidated-Vultee Aircraft on Jan. 23.

Aug. 7, 1951   A Navy Viking 7 rocket sets an altitude record for single-stage rockets, climbing to 136 miles and reaching a speed of 4,100 mph at White Sands.

Oct. 29, 1951   Firing of sixty-sixth V-2 at White Sands Proving Ground concludes U.S. use of these German missiles in upper-atmosphere rocket research.

Feb. 15, 1952   Flight Determination Laboratory is established at White Sands Proving Ground.

July 22, 1952   First production-line Nike makes successful flight.

Aug. 20, 1953   The first successful firing of the Army's Redstone missile is achieved by Redstone Arsenal (RSA) personnel at Cape Canaveral.

Dec. 31, 1954   U.S. Army Ordnance terminates the Hermes Project, during which there was development of the highest performance liquid-fuel rocket in the U.S. to that date; development of the largest solid-propellant rocket motor flown to that date; and development of the first stabilized platform inertial guidance system with air-bearing gyros and accelerometers for ballistic missiles.

July 29, 1955   President Eisenhower endorses USNC-IGY earth satellite proposal. The White House announces that "the President has approved plans . . . for going ahead with the launching of small unmanned earth-circling satellites as part of the U.S. participation in the International Geophysical Year to take place between July 1957 and December 1958. . . ."

Sept. 9, 1955   Project Vanguard is approved and the program placed under Navy management and Dept. of Defense monitorship. Objectives were: to develop and procure a satellite launching vehicle, to place at least one satellite in orbit around the earth during the IGY, to accomplish one scientific experiment, and to track flight.

Feb. 1, 1956
Army Ballistic Missile Agency (ABMA) is activated. ABMA's nucleus was Redstone Arsenal's Guided Missile Development Division.

Mar. 14, 1956
ABMA launches the first Jupiter A (modified Army Redstone missile) at Cape Canaveral, Florida.

Sept. 20, 1956
A Jupiter C missile attains an altitude of 680 miles and a range of more than 3,300 miles.

Dec. 8, 1956
First test rocket in the U.S. IGY satellite program, a one-stage NRL Viking, attains an altitude of 126 miles and a speed of 4,000 mph. The rocket carried a "minitrack" radio transmitter, ejected at 50 miles and tracked.

May 31, 1957
The 1500-mile Jupiter is fired successfully at Atlantic Missile Range. This was the Army's first successful launch of an IRBM for this distance.

Oct. 4, 1957
Sputnik I, first earth satellite, is launched by the USSR and remains in orbit until January 4, 1958.

Nov. 3, 1957
Sputnik II, carrying a dog named Laika, is launched by the USSR. The satellite remained in orbit until April 14, 1958.

Dec. 17, 1957
First successful firing of USAF Atlas ICBM, the missile landing in the target area after a flight of some 500 miles, on the fifty-fourth anniversary of the Wright brothers' first flight.

Jan. 13, 1958
Explorer I, the free world's first earth satellite is placed in orbit by a Juno I (modified Jupiter C), its payload discovering the radiation belt identified by Dr. James A. Van Allen.

July 29, 1958
President Eisenhower signs the National Aeronautics and Space Act, redefining the U.S. space program.

Oct. 1, 1958
The Navy's Project Vanguard and 400 Naval Research Laboratory scientists are assigned to NASA on its first day of operation. Also transferred to NASA from the Department of Defense were lunar probes, satellite projects, and engine-development research programs.

Jan. 2, 1959
USSR launches Lunik I into a solar orbit, with a reported total weight of 3,245 pounds. Lunik I, called

Metchtá (dream) by the Russians, was the first man-made object placed in orbit around the sun.

April 2, 1959
Seven astronauts are selected for Project Mercury: Captains L. Gordon Cooper, Jr., Virgil I. Grissom and Donald K. Slayton, USAF; Lt. N. Scott Carpenter, Lt. Cmdr. Alan B. Shepard, Jr., Walter M. Schirra, Jr., USN; and Lt. Col. John H. Glenn, USMC.

Sept. 9, 1959
First launch of operational Air Force Atlas ICBM from Vandenberg Air Force Base is successful. Second Atlas ICBM is fired from AMR the same day.

Sept. 12, 1959
Russia's Lunik II, launched with a total payload weight of 858.4 pounds, becomes the first man-made object to hit the moon (September 13). This launching coincided with the visit of Premier Nikita Khrushchev to the United States.

Mar. 11, 1960
Pioneer 5 is launched to measure radiation and magnetic fields between Earth and Venus.

April 1, 1960
Tiros 1, first known weather-observation satellite, is launched and takes pictures of Earth's cloud cover from altitude of 450 miles.

July 29, 1960
Project Apollo, advanced manned-spacecraft program, is announced.

May 5, 1961
Freedom 7, manned Mercury spacecraft carrying Astronaut Alan B. Shepard, Jr., as pilot is launched at Cape Canaveral as first American manned space flight. Flight lasted 14.8 minutes and reached an altitude of 115 miles.

Feb. 20, 1962
Mercury spacecraft Friendship 7, with Lt. Col. John H. Glenn as astronaut, is launched on first U.S. manned orbital space flight, covering 81,000 miles in 4 hours and 56 minutes.

Aug. 27, 1962
Mariner 2 is launched from Cape Canaveral on 180-million-mile four-month flight to Venus, later transmitting first data from that planet's vicinity.

Oct. 25, 1962
First live two-way radio broadcast is conducted via Telestar.

June 12, 1963
Project Mercury is officially ended, having achieved its goals.

July 26, 1963
Syncom 2 communications satellite is put in orbit, producing telephone, teletype, and photo facsimile communications between U.S. and Africa.

July 20, 1964
Space flight of Sert 1 spacecraft marks first successful operation in space of an electric rocket engine.

July 28, 1964
Ranger VII spacecraft is launched from Cape Kennedy (formerly Cape Canaveral) on its way to the Moon, sending back 4,316 clear photos before impact on lunar surface.

Mar. 23, 1965
Gemini 3 spacecraft, with Astronauts Virgil I. Grissom and John W. Young on board, is launched and makes three orbits in 4 hours and 53 minutes, the first time a manned spacecraft was maneuvered in orbit.

April 6, 1965
Intelsat 1, first commercial communications satellite, is placed in synchronous equatorial orbit above Atlantic Ocean.

June 3, 1965
Gemini 4 spacecraft, with Astronauts James A. McDivitt and Edward H. White II as pilots, launched. It made 62 revolutions around the Earth in 97 hours and 56 minutes, during which White became the first American to walk in space.

Nov. 28, 1964
Mariner IV is launched on 228-day, 325-million-mile flight to Mars, sending back first close photos of that planet.

Dec. 4, 1965
Gemini 7, piloted by Astronauts Frank Borman and James A. Lovell, Jr., is launched on 14-day mission, the longest U.S. flight to date, lasting 330 hours and 35 minutes. Eleven days after launch, the spacecraft achieved its historic rendezvous in orbit with Gemini 6, launched December 15 and piloted by Astronauts Walter M. Schirra, Jr., and Thomas P. Stafford.

May 30, 1966
Surveyor I is launched, to become first U.S. spacecraft to make a soft landing on Moon.

July 18, 1966
Gemini 10, eighth manned flight in Gemini series, is launched on successful rendezvous and docking mission. Astronauts John W. Young and Michael Collins were pilots and performed first docked-spacecraft maneuver, also rendezvousing with Gemini 8's target.

Aug. 10, 1966 — Lunar Orbiter I, unmanned, is launched, to become first U.S. spacecraft to enter lunar orbit, taking first pictures of Earth from vicinity of Moon. Orbiter intentionally crashed on the far side of the Moon.

Nov. 11, 1966 — Gemini 12, last of the series, is launched with Astronauts James A. Lovell, Jr., and Edwin E. Aldrin, Jr. Mission lasted nearly four days.

Jan. 27, 1967 — Three-man crew training for first Apollo flight—Virgil I. Grissom, Edward H. White II, and Roger B. Chafee—die when flash fire sweeps through Command Module at Kennedy Space Center.

Jan. 6, 1968 — Surveyor VII is launched to soft land in Moon's highlands, return TV pictures, perform lunar digging experiments, and detect laser beams directed from Earth.

Dec. 21, 1968 — Apollo 8, second manned mission in program and first to orbit Moon, is launched. Astronauts: Frank Borman, James A. Lovell, Jr., and William A. Anders. Mission duration: 6 days, 3 hours. Twenty hours in lunar orbit. Circled Moon 10 times. Support facilities tested. Photographs taken of Earth and Moon. Live TV broadcasts.

May 18, 1969 — Apollo 10 is launched for dress rehearsal of Moon landing. Astronauts: Eugene A. Cernan, John W. Young, and Thomas P. Stafford. Mission duration: 8 days, 3 minutes. First manned CSM/LEM operations; simulation of first lunar landing profile. In lunar orbit 61.6 hours, with 31 orbits. LEM taken to within 47,000 feet of lunar surface. First live color TV from space. LEM ascent stage jettisoned in orbit.

July 16, 1969 — Apollo 11 is launched. Man's first landing on the moon. Neil A. Armstrong, commander; Michael Collins, CM pilot; Edwin E. Aldrin, Jr., LEM pilot. Liftoff occurred at 8:32 A.M., July 16. LEM touchdown on moon 4:17:43 P.M., July 20. Neil Armstrong took man's first step on the moon 10:56:15 P.M., July 20. Edwin Aldrin stepped on the lunar surface at 11:15:15 P.M.

July 29, 1969 — Mariner 6 takes first pictures of Mars, received at JPL.

Nov. 14, 1969   Apollo 12 is launched. Astronauts: Charles Conrad, Jr., Alan L. Bean, and Richard F. Gordon. Lunar Module lands in Sea of Storms.

Mar. 31, 1970   Explorer I, first U.S. satellite, launched Jan. 31, 1958, reenters Earth atmosphere after completing 58,408 revolutions and traveling 2.67 billion miles.

April 11, 1970   Apollo 13, third lunar landing mission attempt, is launched. Astronauts: James A. Lovell, Jr., John L. Swigert, Jr., and Fred W. Haise, Jr. Mission duration: 5 days, 22.9 hours. Mission aborted after rupture of Service Module oxygen tank. Classed as "successful failure" because of experience in rescuing crew.

Dec. 2, 1970   Explorer XLII is launched into equatorial orbit from the San Marco platform off the coast of Kenya by an Italian crew, the first American spacecraft to be sent aloft by men of another country.

Feb. 2, 1971   Apollo 14, third successful lunar landing mission, is launched. Astronauts: Alan B. Shepard, Jr., Stuart A. Roosa and Edgar D. Mitchell. Lunar Module lands in Frau Mauro area.

July 26, 1971   Apollo 15, fourth successful lunar landing mission and first of "J" series using Lunar Roving Vehicle (LRV), is launched. Astronauts: David R. Scott, James B. Irwin and Alfred M. Worden. Lunar Module lands in Hadley Rille region near Apennine Mountains.

Nov. 3, 1971   Mariner 9, launched May 30, goes into orbit of Mars, first to circle another planet, took first close photos of Mars' moons, Deimos and Phobos, and of a Mars dust storm.

Mar. 2, 1972   Pioneer 10 is launched from Cape Kennedy. The spacecraft's trajectory intersected Mars orbit on May 25, 1972, and entered the asteroid belt on July 15, 1972. On February 2, 1973, the spacecraft emerged safely from the asteroid belt and continued on to Jupiter. The Jovian gravity accelerated the spacecraft toward Saturn.

April 16, 1972   Apollo 16, fifth successful lunar landing mission is launched. Astronauts: John W. Young, Thomas K. Mattingly II and Charles M. Duke, Jr. Landed in Descartes, highlands area. Mission duration 11 days,

1 hour, 51 minutes. First study of highlands area. Selected surface experiments deployed. Tests started by Apollo 15 crew continued. Ultraviolet camera/spectrograph used second time. In lunar orbit 126 hours, with 64 orbits. Two hundred and thirteen pounds of material gathered.

Dec. 7, 1972   Apollo 17, sixth lunar landing and final mission in Apollo series, launched. Astronauts: Eugene A. Cernan, Ronald B. Evans, and Harrison H. Schmitt, the last-named the first geologist on the Moon. Landed in Tauris-Littrow area. In lunar orbit 147 hours, 48 minutes. Two hundred and twenty-five pounds of material gathered.

April 6, 1973   Pioneer 11 launched from Cape Kennedy toward planet Jupiter.

May 14, 1973   Skylab 1, the nation's first orbiting laboratory, is launched with a three-man crew, Charles Conrad, Jr., Joseph P. Kerwin, and Paul J. Weitz, scheduled to rendezvous with it the next day. Twelve minutes after launch, signals revealed the meteoroid shield had been torn away and one of two solar panels only partially deployed.

May 25, 1973   After eleven days during which technicians developed a parasol as a substitute for the heat shield and tools with which astronauts could cut the metal strap keeping the solar wing from deploying, the crew was launched and rendezvoused with Skylab. A 28-day mission was completed on June 22.

July 28, 1973   The second Skylab crew is launched, consisting of Alan L. Bean, Jr., Jack R. Lousma, and Owen K. Garriott. En route, a rocket thruster on the spaceship began leaking. On August 2, a second thruster developed leaks, leaving only two in operable condition. The astronauts completed a 59-day mission and returned to Earth via their disabled craft on September 25, 1973.

Nov. 16 1973   Third Skylab is launched. Crew, Gerald P. Carr, William R. Pogue, and Edward Gibson spent 84 days, 1 hour and 19 minutes in the space station. More than 1500 hours were spent in numerous scientific observations, such as solar astronomy, earth observations, astrophysics, and material science, among others, including more than 150 hours observing the comet Kohoutek. The astronauts returned to earth on February 8, 1974.

Mar. 20, 1974   Pioneer 11 emerged undamaged in its flight through the asteroid belt on its way toward Jupiter.

April 19, 1974   Pioneer 11's trajectory adjusted to have it continue on to Saturn after flying across orbit of Saturn.

Selected Bibliography

Aikens, David S. *John Glenn, First American in Orbit.* Huntsville, Alabama: Strode, 1969.

Astronauts Themselves, The. *We Seven.* New York: Simon & Schuster, 1962.

Braun, Wernher von and Ordway, Frederick I., III. *History of Rocketry and Space Travel.* New York: Crowell, 1966.

Coombs, Charles. *Rocket Pioneer.* New York: Harper, 1965.

Ford, Brian. *German Secret Weapons—Blueprint for Mars.* New York: Ballantine, 1969.

Freedman, Russell. *2000 Years of Space Travel.* New York: Holiday House, 1963.

Goodrum, John C. *Wernher von Braun, Space Pioneer.* Huntsville, Alabama: Strode, 1969.

Ley, Willy. *Missiles & Men in Space.* New York: New American Library, 1969.

———. *Rockets, Missiles & Men in Space.* New York: Viking, 1968.

Newlon, Clarke. *Famous Pioneers in Space.* New York: Dodd, Mead, 1963.

Sharpe, Mitchell R. *Yuri Gagarin, First Man in Space.* Huntsville, Alabama: Strode, 1969.

Stoiko, Michael. *Project Gemini: Step to the Moon.* New York: Holt, Rinehart and Winston, 1963.

————. *Soviet Rocketry: Past, Present and Future.* New York: Holt, Rinehart and Winston, 1963.

Titov, Gherman and Caidin, Martin. *I am an Eagle!* New York: Bobbs-Merrill, 1962.

Verral, Charles Spain. *Rocket Genius.* New York: Scholastic Book Services, 1969.

Index

Index